THE LIBRARY
ST. MARY'S COLLEGE OF MARYLAND
ST. MARY'S CITY, MARYLAND 20686

MATHEMATICAL MODELLING METHODOLOGY, MODELS AND MICROS

Mathematics and its Applications
Series Editor: G. M. BELL, Professor of Mathematics, King's College (KQC), University of London

Statistics and Operational Research
Editor: B. W. CONOLLY, Professor of Operational Research, Queen Mary College, University of London

Mathematics and its applications are now awe-inspiring in their scope, variety and depth. Not only is there rapid growth in pure mathematics and its applications to the traditional fields of the physical sciences, engineering and statistics, but new fields of application are emerging in biology, ecology and social organisation. The user of mathematics must assimilate subtle new techniques and also learn to handle the great power of the computer efficiently and economically.

The need of clear, concise and authoritative texts is thus greater than ever and our series will endeavour to supply this need. It aims to be comprehensive and yet flexible. Works surveying recent research will introduce new areas and up-to-date mathematical methods. Undergraduate texts on established topics will stimulate student interest by including applications relevant at the present day. The series will also include selected volumes of lecture notes which will enable certain important topics to be presented earlier than would otherwise be possible.

In all these ways it is hoped to render a valuable service to those who learn, teach, develop and use mathematics. *For full series list see end of book.*

MATHEMATICAL MODELLING METHODOLOGY, MODELS AND MICROS

J. S. BERRY, B.Sc., Ph.D.
Lecturer in Applied Mathematics, The Open University
D. N. BURGHES, B.Sc., Ph.D.
School of Education, University of Exeter
I. D. HUNTLEY, B.A., Ph.D.
Department of Mathematics, Sheffield City Polytechnic
D. J. G. JAMES, B.Sc. (Maths), B.Sc.(Chemistry), Ph.D., F.I.M.A.
Department of Mathematics, Coventry (Lanchester) Polytechnic
A. O. MOSCARDINI, , B.Sc., M.Sc., Dip.Ed.
Department of Mathematics & Statistics, Sunderland Polytechnic

ELLIS HORWOOD LIMITED
Publishers · Chichester

Halsted Press: a division of
JOHN WILEY & SONS
New York · Chichester · Brisbane · Toronto

First published in 1986 by
ELLIS HORWOOD LIMITED
Market Cross House, Cooper Street,
Chichester, West Sussex, PO19 1EB, England
The publisher's colophon is reproduced from James Gillison's drawing of the ancient Market Cross, Chichester.

Distributors:

Australia and New Zealand:
JACARANDA WILEY LIMITED
GPO Box 859, Brisbane, Queensland 4001, Australia

Canada:
JOHN WILEY & SONS CANADA LIMITED
22 Worcester Road, Rexdale, Ontario, Canada

Europe and Africa:
JOHN WILEY & SONS LIMITED
Baffins Lane, Chichester, West Sussex, England

North and South America and the rest of the world:
Halsted Press: a division of
JOHN WILEY & SONS
605 Third Avenue, New York, NY 10158, USA

© 1986 J. S. Berry, D. N. Burghes, I. D. Huntely, D. J. G. James, and A. O. Moscardini/Ellis Horwood Limited

British Library Cataloguing in Publication Data
Mathematical modelling methodology, models and micros. —
(Mathematics and its applications
1. Mathematical models
I. Berry, John, 1947– . II. Series
511'.8   QA401

ISBN 0–7458–0080–7 (Ellis Horwood Limited)
ISBN 0–470–20717–5 (Halsted Press)

Printed in Great Britain

COPYRIGHT NOTICE
All Rights Reserved. No part of this publication may be reproduced, stored in a retrieval system, or transmitted, in any form or by any means, electronic, mechanical, photo-copying, recording or otherwise, without the permission of Ellis Horwood Limited, Market Cross House, Cooper Street, Chichester, West Sussex, England.

Table of Contents

PREFACE 9

Chapter 1. MATHEMATICAL MODELLING—ARE WE HEADING IN THE RIGHT DIRECTION? 11
D. N. Burghes, University of Exeter, UK

Section A. Methodology
Chapter 2. MATHEMATICAL MODELLING OF EDUCATIONAL DATA 27
G. Hanna, The Ontario Institute for Studies in Education, Canada

Chapter 3. MODELLING IN CALCULUS INSTRUCTION—EMPIRICAL RESEARCH TOWARDS AN APPROPRIATE INTRODUCTION OF CONCEPTS 36
G. Kaiser-Mebmer, Gesamthochschule, Kassel, FRG

Chapter 4. ASSESSMENT IN MATHEMATICAL MODELLING 48
K. H. Oke, South Bank Polytechnic, London and A. C. Bajpai, Loughborough University of Technology, UK

Chapter 5. FORMULATION–SOLUTION PROCESSES IN MATHEMATICAL MODELLING 61
K. H. Oke, South Bank Polytechnic, London and A. C. Bajpai, Loughborough University of Technology, UK

Chapter 6. A NEW APPROACH TO MODEL FORMULATION 80
D. E. Prior, Sunderland Polytechnic, UK

Chapter 7. GROUP PROJECTS IN MATHEMATICAL MODELLING 90
G. L. Slater, Sheffield City Polytechnic, UK

Chapter 8. MATHEMATICAL MODELS IN THE TEACHING OF SYSTEMS THINKING 98
E. F. Wolstenholme, University of Bradford Management Centre, UK

Section B. Models
Chapter 9. MODEL VALIDATION–INTERACTIVE LECTURE DEMONSTRATIONS IN NEWTONIAN MECHANICS 115
C. R. Haines, City University, London and D. Le Masurier, Brighton Polytechnic, UK

Chapter 10. CONTINUOUS AND DISCRETE TECHNIQUES IN MATHEMATICAL MODELLING 130
D. J. G. James and M. A. Wilson, Coventry (Lanchester) Polytechnic, UK

Chapter 11. MECHANICS VIA COSMOLOGY 142
P. T. Landsberg, University of Southampton, UK

Chapter 12. DISCRETE LINEAR SYSTEM MODELLING TECHNIQUES 149
K. V. Lever, University of East Anglia, UK

Chapter 13. IMPROVING DRIVER COMFORT IN MOTOR VEHICLES 165
A. Norcliffe, G. G. Rodgers and M. M. Tomlinson, Sheffield City Polytechnic, UK

Chapter 14. MATHEMATICAL MODELS IN FOREST MANAGEMENT 181
W. J. Reed, University of Victoria, Canada

Chapter 15. CASE STUDIES AND CAL IN ENGINEERING MATHEMATICS 191
S. Townend, Liverpool Polytechnic, UK

Section C. Use of the Microcomputer and Simulation
Chapter 16. THE USE OF MICROS IN EVALUATING AND DISPLAYING THE CHARACTERISTICS OF MODELS USED IN CONTROL THEORY 205
R. V. Aldridge, University of East Anglia, UK

Chapter 17. MECHANICS WITH A MICRO 217
C. E. Beevers, Heriot-Watt University, Edinburgh, UK

Chapter 18. COMPUTER APPLICATIONS OF MODELLING FOR MECHANICAL ENGINEERS 223
G. Beswick and A. S. White, Middlesex Polytechnic, UK

Chapter 19. MATHEMATICAL MODELLING USING DYNAMIC SIMULATION 238
R. R. Clements, University of Bristol, UK

Chapter 20. DISCRETE AND CONTINUOUS SYSTEM MODELLING WITH A MICRO NETWORK 249
I. C. Hendry and I. G. Mackenzie, Robert Gordon Institute of Technology, Aberdeen, UK

Chapter 21. FORMULATION IN MATHEMATICAL MODELLING BY ARTIFICIAL INTELLIGENCE 261
F. R. Hickman, South Bank Polytechnic, London, UK

Chapter 22. USING MODISTAT—A MICRO STATS PACKAGE
FOR ILLUSTRATING MATHEMATICAL PRINCIPLES 287
T. H. Mangles, Plymouth Polytechnic, UK

Chapter 23. 'SOFT' COURSE SIMULATION 301
D. E. Prior, Sunderland Polytechnic, Manchester Polytechnic, UK

Second International Conference on The Teaching of Mathematical Modelling
University of Exeter, 16–19 July 1985

ORGANISING COMMITTEE AND THE PROCEEDINGS EDITORS
J. S. Berry, Plymouth Polytechnic
D. N. Burghes, Exeter University
I. D. Huntley, Sheffield City Polytechnic
D. J. G. James, Coventry (Lanchester) Polytechnic
A. O. Moscardini, Sunderland Polytechnic

ADVISORY PANEL
Professor U. D'Ambrosio, Universidade Estadual de Campinas, Brazil
Professor A. C. Bajpai, University of Technology, Loughborough, UK
Professor H. Burkhardt, Shell Centre for Mathematical Education, University of Nottingham, UK
Professor J. N. Kapur, Indian Institute of Technology, Kanpur, India
Sir Harry Pitt, formerly Vice Chancellor University of Reading, UK
Dr H. O. Pollack, Bell Laboratories, New Jersey, USA
Dr A. G. Shannon, The New South Wales Institute of Technology, Sydney, Australia
Professor H. E. Steiner, Institute fur Didaktik der Mathematik, Universität Bielefeld, West Germany

Preface

Over the past two decades, there has been a growing awareness that mathematics should not be taught in isolation from its applications, and consequently, many teachers and lecturers today incorporate applications and mathematical modelling into their existing courses.

In Section A, several authors review the present methodology in the process of problem solving and propose new ideas in the formulation stage of mathematical modelling.

Teachers of mathematical modelling are always looking for fresh ideas and, accordingly, Section B provides real-life examples and case studies for use in the classroom.

Micros play an important role in most subjects in the classroom and this is equally true in mathematical modelling courses. Section C offers ideas in this area from several experienced teachers.

By reading this text, we hope that you will find useful and practical information which will enhance the study of mathematics both for students new to the subject and for experienced teachers. A subsequent and successor text *Mathematical Modelling Courses* will provide advice and guidance for those proposing to set up mathematical modelling courses.

Acknowledgement

The Organising Committee would like to express their thanks to the following bodies for financial and other aid towards the running of the conference:

Cham Limited
Coventry (Lanchester) Polytechnic
Exeter University
National Westminster Bank (Exeter)
The Open University
Sheffield City Polytechnic
Sunderland Polytechnic

They would also like to put on record their gratitude to Ann Tylisczuk, Nigel Weaver and particularly to the conference Secretary, Sally Williams.

1

Mathematical Modelling—are we Heading in the Right Direction?

D. Burghes
University of Exeter, UK

---

1. INTRODUCTION

It is only two years since our first international conference, yet there have been many changes. Nearly all our papers this time are very much focused on how we can teach mathematical modelling, whereas in our first conference, we had many papers on lecturers' favourite models, which often have little relevance to other teachers.

If this change in emphasis is being reflected in the way we teach mathematics in school, further and higher education, then we will really be transforming the way we teach and students learn mathematics. In this chapter, I want to review progress that we have made in the teaching of mathematical modelling and attempt to answer the following questions.

(i) What do we mean by mathematical modelling?
(ii) What progress have we made in the teaching of mathematical modelling?
(iii) Do we know where we are going?
(iv) How do we get there?

2. WHAT DO WE MEAN BY MATHEMATICAL MODELLING?

This is not an easy question to answer! I am not sure that we could find complete agreement amongst our participants. I like to use a diagram of

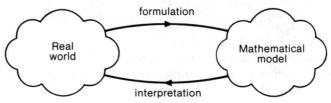

the type above to describe the important features of modelling—but not all mathematical modelling can fit neatly into this form. Nevertheless it does bring home what is to me the essential feature—namely that of using mathematics to solve real problems.

How real the problems are which can be used in modelling courses is open to debate. There are few totally real problems but there are some that are very close to reality. For example, the problem

> Design a £1 book of stamps which caters for 1st and 2nd class post in Great Britain

is almost reality—the Post Office is planning to bring in £1 coin automatic machines. Unfortunately we do not know the complete background though—we would need to know

(i) What do customers buying these books of stamps actually want?
(ii) Does the Post Office deliberately put in 1p and 2ps, which are not usually of use, because they represent profit?
(iii) The percentage of 1st class to 2nd class post sent.
(iv) How often will the postage rates change?

So, in fact, the seemingly easily stated problem can soon become a project, which if used in a teaching situation could involve a survey of customers. We are clearly dealing with reality—although we can only rarely make the problem totally real.

As another example, consider the grid below:

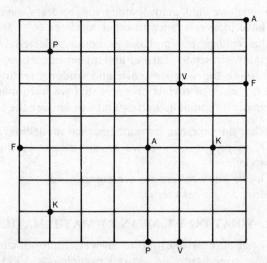

The problem here is to find a connection between corresponding letters (i.e. A → A, F → F, K → K, P → P, V → V) with the conditions

(i) Only use the grid lines.
(ii) No two lines must cross or touch at a point or coincide.
(iii) No line must pass through a different letter (e.g. A to A must not pass through V).

It might take you a while to find the solution—indeed, I call this a 'eureka' investigation since there is a breakthrough which must be made before success is found. (The solution is given in Appendix 1—but have a good try before resorting to looking up the solution.) I am sure that we can agree that this is a mathematical investigation, but is it modelling? Not really, in its present form, but we can make this sort of problem nearer reality by looking at the circuit diagram in Fig. 1 and asking the question 'Can it be

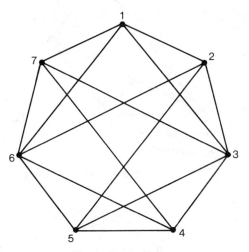

Fig. 1. Circuit diagram

redrawn with no crossovers?' Here the vertices 1–7 are fixed and the connections can be regarded as flexible wires which can be moved. This type of problem is central to efficient design of silicon chips where crossovers have to be insulated, with consequent increase in cost. The solution to this problem is relatively straightforward—a possible configuration with no crossovers is shown in Fig. 2. We are still left with the question 'Is this modelling?' Again, not really—but it is certainly mathematical and it is connected with reality! Indeed, the real problem is the design of an algorithm to cope with problems of this type. You might like to try out your method on the circuit diagram in Fig. 3—it can be redrawn without crossovers (see Appendix 2).

An algorithm for solving such problems is given in the Open University course TM361, Graphs, Networks and Design—Unit 12.

Although these examples are *not* modelling in the restricted sense, they are problems, dealing with reality and needing mathematics to help in finding solutions—so surely we would not want to discount them from modelling courses. They are all about *strategy* which is definitely one of the most important concepts in effective modelling, and we should be aiming to improve this skill.

So perhaps we should not be asking this question; let us agree to accept that definitions of mathematical modelling will vary—but that this does not matter. What is important though, is that students on modelling

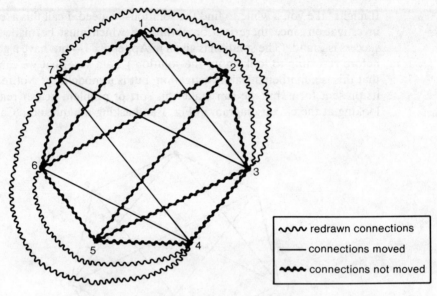

Fig. 2. Redrawn circuit diagram.

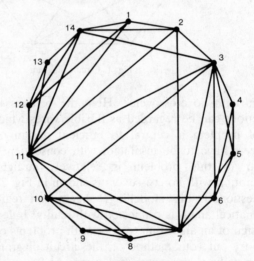

Fig. 3. Circuit diagram

courses realise that they are trying to use mathematics in solving problems connected with reality, and as teachers our aim should be to motivate and improve the skills required.

3. WHAT PROGRESS HAVE WE MADE?

If we look at the many varied papers given at this conference, it is clear that we have made progress—there are many tutors who have been experimenting with modelling courses over the past few years. In higher education in Great Britain, the Polytechnics have led the way, and most (if not all) polytechnic degrees in mathematics will have courses in modelling, and in some it is the main theme of the applied courses throughout the degree. It is a similar story in many B.Ed. degrees for intending mathematics teachers, where modelling is used as a unifying theme for all applications of mathematics. It is only in universities that similar progress has not been forthcoming. Why is this? There are a number of possible reasons.

(i) Many academics would argue that they are modelling all the time throughout applied courses—so why is a special course needed for modelling?
(ii) Modelling is regarded as not quite academically acceptable—all sincere and ambitious lecturers should get on with the serious business of producing more high level research papers, and not get their hands dirty.
(iii) Lecturers, often brought up exclusively in an academic environment, are frightened of the unknown.

It does seem unfortunate that we have such a small university participation at this conference.

At secondary school level, the emphasis of nearly all the changes now taking place is towards mathematics becoming more practical and relevant. The assessment objectives for the GCSE, due to be examined for the first time in 1988, are given below. Particular note should be made of 3.12/13/15 as well as the extra assessment objectives 3.16/17, which require some form of course-work assessment. These aims are very ambitious, and it remains to be seen how effective they will be in practice.

Assessment objectives

The objectives which follow set out essential mathematical processes in which candidates' attainment will be assessed. They form a minimum list of qualities, abilities and skills. The weight attached to each of these objectives may vary for different levels of assessment within a differentiated system.

Any scheme of assessment will test the ability of candidates to carry out the following.

3.1 Recall, apply and interpret mathematical knowledge in the context of everyday situations.
3.2 Set out mathematical work, including the solution of problems, in a logical and clear form using appropriate symbols and terminology.
3.3 Organise, interpret and present information accurately in written, tabular, graphical and diagrammatic forms.
3.4 Perform calculations by suitable methods.
3.5 Use an electronic calculator.
3.6 Understand systems of measurement in everyday use and make use of them in the solution of problems.
3.7 Estimate, approximate and work to degrees of accuracy appropriate to the context.

3.8 Use mathematical and other instruments to measure and to draw to an acceptable degree of accuracy.
3.9 Recognise patterns and structures in a variety of situations, and form generalisations.
3.10 Interpret, transform and make appropriate use of mathematical statements expressed in words or symbols.
3.11 Recognise and use spatial relationships in two and three dimensions, particularly in solving problems.
3.12 Analyse a problem, select a suitable strategy and apply an appropriate technique to obtain its solution.
3.13 Apply combinations of mathematical skills and techniques in problem solving.
3.14 Make logical deductions from given mathematical data.
3.15 Respond to a problem relating to a relatively unstructured situation by translating it into an appropriately structured form.

Two further assessment objectives can be fully realised only by assessing work carried out by candidates in addition to time-limited written examinations. From 1988 to 1990 all Examining Groups must provide at least one scheme which includes some elements of these two objectives. From 1991 these objectives must be realised fully in all schemes.

3.16 Respond orally to questions about mathematics, discuss mathematical ideas and carry out mental calculations.
3.17 Carry out practical and investigational work, and undertake extended pieces of work.

Also, the draft national grade criteria document, recently published by the Schools Examinations Council, mentions mathematical modelling as one of a number of topics which must be mastered by candidates passing at the upper grades.

4. DO WE KNOW WHERE WE ARE GOING?

So the past few years has seen much progress on the theme of teaching mathematics through its applications, with particular emphasis on mathematical modelling. There are, however, still many problems in mathematics education and it is not suggested that modelling is going to solve them all. One associated problem is the *assessment* of modelling. There are many possible forms of assessment, for example:

  (i) Close book examinations
 (ii) Mathematics practicals
(iii) Projects
 (iv) Continuous assessment
  (v) Essays
 (vi) Group assessment
(vii) No assessment

I have no doubt that my most effective modelling courses have been given using (vii)—no assessment. The courses were for teachers who were well motivated, and assessment would have been a real nuisance—indeed, it

Mathematical Modelling—Heading in the Right Direction

would undoubtedly have done educational harm. I realise though that the ideal situation will not usually exist—students will not be so well motivated, and anyway CNAA, for example, will insist on some form of assessment.

One of the most important benefits, to my mind, of modelling courses, is the interaction between students, working in groups. Much of mathematics does not require interaction, but modelling gives a welcome opportunity for group work. This will also be true of course-work assignments for the new GCSE. The difficulty will again be that of assessment—how do you assess contributions of different members of a team? Indeed, is it fair to try to do so?

So, although I welcome the move that we are making, and I look forward to future experiments, I think that we should exercise caution and attempt to assess the developments that we are making.

5. HOW DO WE GET THERE?

This is difficult to answer, since I am not sure where we are going. One thing though that I am convinced about—staff involved in teaching modelling courses should either

(i) have direct experience in applying maths in practical problem-solving situations

or

(ii) have been involved in substantial in-service work, which can give simulated experience in modelling

I think it is a recipe for disaster if tutors, without the relevant background and experience, have to teach modelling courses. Now that we have collectively more experience in teaching modelling courses, it is time, both at school and higher education, to provide intensive in-service courses in modelling. The experience gained on such courses must be a *minimum* requirement before embarking on teaching a modelling course.

As an example of the sort of simulated modelling experience that can be given on in-service courses, try the following problem:

> Design a method of assessing the reading age of an article and use it to find the reading age of the following four articles

Is today's single girl a free-wheeling, free-loving lady like Connie in the TV series? Or is she really a lonely person who longs to find a partner and settle down?
There are 1,750,000 single women in Britain today. To find out the truth about their way of life the News of the World conducted an intensive survey among those aged from 20 to 44.

And our researchers have come up with some astonishing facts about sex and the single girl.

A hefty percentage, nearly 20 per cent, of single girls said they DIDN'T want a permanent relationship at present.

Women who have had many lovers are not so keen to settle into wedded bliss. There was one swinger in every 100 singles who preferred one night stands or a series of partners to any kind of committed relationship.

And living together permanently without marriage is surprisingly unpopular. Less than one in five would be happy with 'living in sin.'

Yet 65 per cent of women would be happy to live with a boyfriend BEFORE marriage.

Indeed, 18 per cent of those in our survey had already done so at one time, the highest numbers of those 'trying him out' being Southern middle-class girls.

Girls from the Midlands were least likely to consider a live-in lover.

After Mr Gromyko

If nothing else the apppointment of Mr Andrei Gromyko as Soviet President is the end of an era of sorts. The Methuselah of world diplomacy—he was casting vetoes in the United Nations Security Council earning for himself the nickname of 'Old Stone Face' when Ronald Reagan was still busy making movies—has been shunted sideways to fulfil a largely ceremonial, however necessary, public relations exercise in the Kremlin. After spending nearly 30 years fairly successfully reconciling the national interests of the Soviet Union with its professed goal of world revolution, Mr Gromyko now has to settle for a bit of peace and quiet and the respect and status (but little power) a Soviet presidency usually entails.

Recovering, Western leaders might well be asking what happens next in Russian foreign policy. On the surface the appointment as Foreign Minister of a charming Georgian party boss, and three-star police general, without any foreign affairs experience apart from the odd trip to a Third World country, is an exercise in Russian cynicism. It could also be interpreted as another example of Mr Mikhail Gorbachev's desperate house-cleaning—like other sectors of the Russian bureaucracy the Foreign Ministry has become set in its ways and even corrupt. Yet ultimately, one suspects, we are about to see a change of style, not of substance, in Russian foreign policy with Mr Gorbachev, as befits a Russian leader, doing all the essential globe-trotting, talking and negotiating, a Nikita Khrushchev of the 1980s and just possibly into the next century.

It should be noted that Mr Gorbachev has already shown he knows a thing or two about familiar hardline anti-Western rhetoric, and there is precious little evidence of any slowdown in the Russian arms build-up. To take just one example, his 'moratorium' on SS-20 missile deployment has already been exposed as a sham. The Soviet Union's foreign policy, like that of any modern State, is conducted in the interests of the State as perceived by the likes of Mr Gorbachev, his Politburo and his military, albeit from time to time distorted by the lens of Marxist-Leninism. A debate is already taking place in the United States over prospects for the success of the meeting in Geneva in November between Mr Gorbachev and Mr Reagan. It is a long way ahead but it is safe to say that the outlook for an improvement in East–West relations remains stony.

We'll carry on until the scabs are 6ft under

The bridge leaving Walmer is known as Scab Bridge because of the hate message daubed on it.

A mile down the road, pubs like the Yew Tree on the Mill Hill estate are no-go areas to miners who worked.

A regular there said: 'If one, just one, of those bastards puts a foot in here he'll have his teeth knocked down his throat.'

Another no-go area is the Kent Miners' Welfare Club, where a man on the door spelled out the latest club rule.

'No scabs,' he snapped. 'Those bloody traitors are the last people on earth who'd be welcome here.'

Pit safety worker Peter Brindley, a veteran of the 1974 strike which toppled the Heath Government—in which he was the first Kent picket arrested—declared: 'Our lads are being soft on the scabs. You can't do enough to them. It will continue until they're all six feet under.'

Peter, 35, and his wife Mary, 38, still feel bitter about the way their middle-class neighbours treated them during the hard, lean months of the strike.

He said: 'Not one of them offered us help. We didn't even get a few words of encouragement.'

9. Hyena and Hare

Hyena had no water. He said to himself, 'Now I must dig a well.' So off he went with his spade over his shoulder, and on his way he met neighbour Hare.

'Good day, neighbour Hare!'

'Good day, neighbour Hyena! Where are you going with your spade over your shoulder?'

'I'm going to dig a well, because I have no water.'

Hare said, 'I have no water, either.'

Hyena said, 'Then come and help me to dig, and we will share the well water.'

Hare said, 'What do I want with a well? If I am thirsty in the morning, I drink the dew on the grass. If I am thirsty in the evening—there is the dew again.'

Hyena said, 'Then you won't help me dig?'

Hare said, 'No, I am busy about other matters.'

So Hyena dug his well all by himself. And early next morning he took a bucket and went to his well to fetch water. What did he see? He saw Hare's footprints all round the well.

'Oh ho!' said he, 'Little Liar Hare! You said you only drank the dew, but now you have been drinking my water! For that I will punish you!'

Hyena ran home; he fetched his axe and his saw. He cut down a tree branch, and out of the branch he made a big doll. He covered the doll all over with sticky tar. Now she looked just like a little black girl. It took him all day making that doll; and when evening came, he put her to lean over the well. Then he went home to bed.

In the night the moon was shining among clouds. Now see—someone coming to the well: *tipperty-tip*, *tipperty-tip*, Hare coming with his bucket to fetch water. What does he see by the dim light of the moon? He sees a little black girl standing to look down into the well. No, no, that won't do! The little black girl may go and tell Hyena about Hare taking water from his well. So Hare says, 'Little black girl, you're trespassing! Go away!'

Little black girl doesn't move, doesn't speak.

Hare gets angry. 'Little black girl, go away, or I shall smack you on the neck!'

Little black girl doesn't move.

Hare lifts his hand. *Smack!*

What happens? His hand sticks to the tar.

'Little black girl, let go my hand, or I shall smack you over the eyes with my other hand.'

Little black girl doesn't move.

Hare lifts his other hand. *Smack!* His other hand sticks.

'Little black girl, let go, let go! Little black girl, do you see this foot? If you don't let go I shall kick you! I shall kick you so hard that you'll think a horse had lashed out at you!'

(Possible solutions are given in Appendix 3.)

As well as providing courses, it is time we organised ourselves and provided an up-to-date database of possible problems and projects for modelling courses. Teachers need a constant flow of resource material, and it is time to use today's technology to provide us with the resources needed. For the 11–16 age range, the Department of Trade and Industry is considering the introduction of a national resource database for school teachers. We urgently need a similar initiative in higher education for mathematics (and other subjects?). It could not only include suggestions for projects, but also relevant data and information.

So here is a plea for: (i) more in-service courses and (ii) a database for suitable resources. I hope that the TAM group can help in taking up these suggestions.

6. CONCLUDING REMARKS

So why do we want to bring in modelling courses? The simple answer is *motivation*—and not just motivation for our students, but self-motivation. We want to believe that mathematics is an important useful tool, and this is our attempt to show this.

Whilst there are good reasons for introducing courses which teach our students: how to

(i) apply mathematical analysis to practical situations,
(ii) improve their attitudes and strategy to problem solving,
(iii) develop their adaptability to cope with changing situations.

I still believe that the most important reason is that of providing tutors with the motivation that they need. A *well motivated* and *enthusiastic* teacher solves all the problems in mathematics education at a stroke!

APPENDIX 1: SOLUTION TO GRID PROBLEM

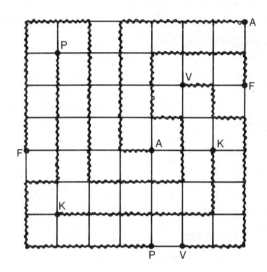

APPENDIX 2

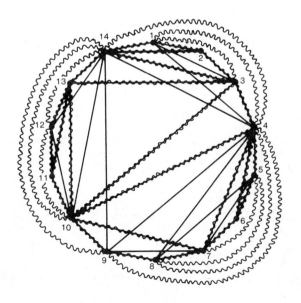

APPENDIX 3: READING AGE MODELS

(i) *Fog index*

$$R = \frac{2}{5}\left(\frac{A}{n} + \frac{100L}{A}\right)$$

where A is the number of words in the passage, n is the number of sentences in the passage and L is the number of words containing three or more syllables (excluding those ending in -ed or -ing).

(ii) *Forecast formula*

$$R = 25 - \frac{N}{10}$$

where N is the number of one-syllable words in a passage of 150 words.

(iii) *Smog formula*

$$R = 8 + \sqrt{p}$$

where p is the number of words with three or more syllables in three passages each of 10 sentences long.

(iv) *Flesch formula*

$$F = 206.835 - 0.846x - 1.015y$$

where x is the exact number of syllables per 100 words and y is the average number of words per sentence, and the figures below indicate the reading age given the F value. You can also use the two outside sides to directly find the reading age. Just join up the x and y values—the intersection on the middle scale gives the reading age.

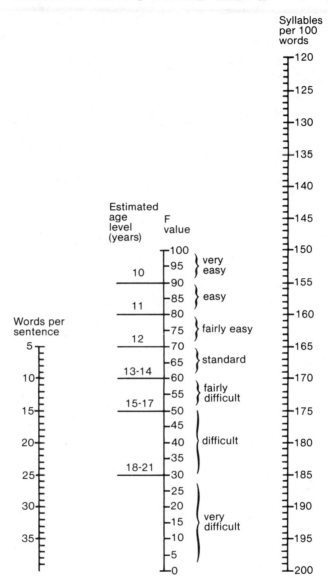

The Flesch formula Reading Ease Score.

Section A
Methodology

2

Mathematical Modelling of Education Data

G. Hanna
The Ontario Institute for Studies in Education, Canada

SUMMARY

The main purpose of this chapter is to explain how a mathematical model incorporating multiple measurements taken on several occasions might be constructed and used to compare two groups of pupils in terms of growth in mathematical ability.

To investigate the structural relationships among successive measurements of the latent variable (mathematical ability as measured by two tests), a series of three-wave two-variable models is developed using techniques drawn from a number of factor-analytic models: the multi-wave multi-variable model, the simultaneous factor-analysis model in several populations, and the structural equation models with structured means. The chapter illustrates the use of the LISREL programme for models with structured means, and in particular demonstrates its usefulness for investigating the structural relationships among the three successive measurements of mathematical ability taken at yearly intervals.

The emphasis is on the methodological issues arising from the application of the structural equation model with structured means to the case of two structural equations, one of which contains two latent variables. The chapter discusses a process of model improvement based upon fitting a series of models, one derived from the other, and comparing alternative models for goodness of fit by examining the differences between their χ^2 values. It also describes a method for estimating the contribution of individual parameters by constructing a series of nested models, each model being more restricted than the previous one in that one additional parameter is constrained.

1. INTRODUCTION

The main purpose of this chapter is to explain how a mathematical model incorporating multiple measurements taken on several occasions might be constructed and used to compare two groups of students, each following a different programme, in terms of growth in mathematical ability from Grade 4 to Grade 6. The chapter illustrates the use of the LISREL program for models with structured means, and also demonstrates its usefulness for investigating the structural relationships among measurements of one true variable taken at yearly intervals.

To investigate the structural relationships among the successive measurements of the true variable postulated in the present study, a series of three-wave two-variable models was developed using techniques drawn from a number of existing factor-analytic models: the multi-wave multi-variable model (Jöreskog, 1979), the simultaneous factor-analysis model in several populations (Jöreskog, 1971), and the structural equation models with structured means (Sörbom, 1974, 1978). The latter models are of particular relevance to this study, though the examples discussed by Sörbom (1978, 1982) all deal with a single structural equation and a single latent covariate, whereas this study uses a model with two structural equations, one of which contains two latent covariates.

Each of the models developed in the present study was tested for goodness of fit by examining the chi-square statistic, the residuals, and the difference in chi-square values between successive models of a series (Jöreskog, 1979; Bentler, 1982).

2. DESCRIPTION OF THE DATA

The data used in this study were drawn from the data pool of the Canadian Bilingual Education Project (Swain and Lapkin, 1981). Data on achievement in English as a first language, French as a second language, and academic subjects were collected in the course of a large-scale evaluation of French Immersion and Regular programmes. Students enrolled in a French Immersion programme get their instruction in French, their second language, while students enrolled in the Regular programme get their instruction in English, their first language. From 1970 to 1979 yearly evaluations were carried out to compare the academic achievement of students in the Immersion programme with that of students in the Regular programme.

For the present study a longitudinal set of data was extracted from the existing pool, encompassing the 144 Immersion students and 59 Regular students for whom there were complete results for the Canadian Test of Basic Skills (CTBS) (Hieronymus et al., 1974) tests in Mathematics Concepts and Mathematics Problems at three successive grade levels (Grades 4, 5 and 6). There are fewer Regular than Immersion students in the longitudinal data because of the focus of the study for which the data was originally collected. When the data were collected the same Immersion students were tracked from year to year, whereas the composition of the Regular groups changed from year to year.

Each student had six tests scores, two at each grade level; all tests were administered in English. The two tests were as follows.

(1) *Mathematics Concepts* A test of about 30 multiple-choice items designed to test how well the student understands the number system and the terms and operations used in mathematics.
(2) *Mathematics Problems* A test of about 30 multiple-choice items designed to assess the student's skills in solving mathematical problems, using single-step or multiple-step problems.

The individual student scores are grade equivalent. In other words, the population mean score for Grade 4 was normalised to 40 at the beginning of the school year, to 45 five months later (the end of January), and to 50 at the end of June (or the beginning of Grade 5). Similarly the mean is 55 in the middle of Grade 5 and 65 in the middle of Grade 6. Since the tests were administered in April, the eighth month of the school year, the mean scores would be expected to be 48, 58, and 68 in Grades 4, 5, and 6, respectively.

3. METHOD

3.1 The measurement model

As shown in Fig. 1, at each grade level the two tests, Mathematics Concepts and Mathematics Problems, are taken to be indicators of the latent variable η mathematical ability; η_4, η_5, and η_6 represent the latent variable mathematical ability in Grades 4, 5 and 6 respectively.

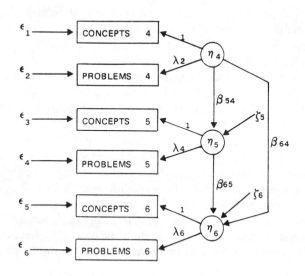

Fig. 1. Model for the measurement of growth in mathematical ability.

The measurement model for mathematical ability is simply

$$\begin{bmatrix} y_5 \\ y_6 \\ y_3 \\ y_4 \\ y_1 \\ y_2 \end{bmatrix} = \begin{bmatrix} 1 & 0 & 0 & v_5 \\ \lambda_6 & 0 & 0 & v_6 \\ 0 & 1 & 0 & v_3 \\ 0 & \lambda_4 & 0 & v_4 \\ 0 & 0 & 1 & v_1 \\ 0 & 0 & \lambda_2 & v_2 \end{bmatrix} \begin{bmatrix} \eta_6 \\ \eta_5 \\ \eta_4 \\ 1 \end{bmatrix} + \begin{bmatrix} \varepsilon_5 \\ \varepsilon_6 \\ \varepsilon_3 \\ \varepsilon_4 \\ \varepsilon_1 \\ \varepsilon_2 \end{bmatrix}$$

The measurement equation assumes that the two tests, Mathematics Concepts and Mathematics Problems, measure a single latent variable on each occasion. The scale for each latent variable was set to be that of the test Mathematics Concepts, by fixing the appropriate factor loading equal to one for both groups. Also the measurement properties of the tests were constrained to be equal across the two groups; that is, they had the same origin of measurement (v_i), the same loadings (λ_j), and the same error variances (ε_i), ($i = 1, 2, \ldots, 6$).

3.2 The structural model

The structural model is represented by the equation:

$$\begin{bmatrix} \eta_6 \\ \eta_5 \\ \eta_4 \\ 1 \end{bmatrix} = \begin{bmatrix} 0 & \beta_{65} & \beta_{64} & 0 \\ 0 & 0 & \beta_{54} & 0 \\ 0 & 0 & 0 & 0 \\ 0 & 0 & 0 & 0 \end{bmatrix} \begin{bmatrix} \eta_6 \\ \eta_5 \\ \eta_4 \\ 1 \end{bmatrix} + \begin{bmatrix} \alpha_6 \\ \alpha_5 \\ \kappa_4 \\ 1 \end{bmatrix} [1] + \begin{bmatrix} \zeta_6 \\ \zeta_5 \\ \eta_4 - \kappa_4 \\ 0 \end{bmatrix}$$

The structural model specifies a causal relationship among the true variables representing mathematical ability at Grades 4, 5, and 6. Specifically, it implies that η_4 has a direct effect on both η_5 and η_6, and that η_5 has an affect only on η_6.

It should be added that a more restricted quasi-simplex model (Jöreskog, 1979), one in which there is no direct effect of η_4 on η_6 (i.e. $\beta_{64} = 0$) but in which there is an indirect effect mediated through η_5, was also postulated.

The origin of measurement and the mean of the latent variable cannot be identified simultaneously, but the differences between the groups can be estimated. When the mean of the latent variable, η_4 is fixed to zero for the Immersion group, the parameter representing the expectation of η_4 for the Regular group is the mean difference in initial mathematical ability between the two groups. The same reasoning applies to η_5 and η_6; only differences between the groups can be estimated. However, it is meaningful to consider these differences as the effect of programme when β parameters (the regression weights) can be shown to be equal in the two groups.

4. PROCEDURE

Following Jöreskog and Sörbom (1981), the present analysis of mean structures was carried out on an augmented moment matrix rather than on

Table 1. Correlations, means and standard deviations. Immersion Group
$N = 144$

Test	C4	P4	C5	P5	C6	P6	Means	Standard deviations
Concepts 4	1						49.20	9.15
Problems 4	0.75	1					48.10	9.53
Concepts 5	0.73	0.68	1				61.33	9.56
Problems 5	0.63	0.64	0.70	1			58.61	10.32
Concepts 6	0.71	0.62	0.73	0.69	1		73.24	11.54
Problems 6	0.60	0.57	0.67	0.64	0.73	1	67.37	11.60

a covariance matrix (i.e. a constant variable 1 is added). The input to the LISREL programme (Tables 1 and 2) consisted of a correlation matrix, a vector of standard deviations, and a vector of means.

The differences between the two groups were investigated by using a series of models. An initial model was first postulated, in which the measurement parameters were constrained to be invariant across the two groups but the structural parameters were left free in both groups. After verifying that this initial model yielded an acceptable fit, a series of nested models was tested, each new model being more restricted than the previous one in that one additional parameter is constrained.

Testing nested models makes it possible to assess the contribution of an individual parameter to the goodness of fit, through the following procedure. If A is a model with x degrees of freedom, a new model B may be formulated such that one additional parameter is either fixed to the value zero or constrained to be invariant across the two groups. B, then, has $x + 1$ degrees of freedom. The new restriction can then be tested by looking at the chi-square difference $(A - B)$ with 1 degree of freedom. If the difference between the chi square for A and that for B does not exceed the critical value at a chosen level, then the new restriction cannot be rejected. If, on the other hand, the chi-square difference is large, then the new restriction imposed may be rejected in favour of the original model (the model with fewer constraints on its parameters).

Table 2. Correlations, means and standard deviations. Regular Group
$N = 59$

Test	C4	P4	C5	P5	C6	P6	Means	Standard deviations
Concepts 4	1						49.24	9.40
Problems 4	0.66	1					46.58	9.30
Concepts 5	0.63	0.52	1				58.47	9.82
Problems 5	0.64	0.66	0.61	1			55.14	10.86
Concepts 6	0.62	0.55	0.72	0.64	1		70.76	12.14
Problems 6	0.49	0.59	0.49	0.61	0.67	1	68.08	10.81

5. RESULTS AND DISCUSSION

The initial model tested in the present study (M1) postulated the equivalence of the measurement properties (equality of factor loadings, equality of the variances of measurement errors) as well as that of the errors in the structural equation, across the two groups. This initial model can be stated more formally as follows: Λ_y, Θ_ε, Ψ and v_i are invariant across the two groups.

Of the 27 parameters to be estimated, 18 are invariant across the two groups: 3 factor loadings, 6 uniquenesses, 3 regression residuals and 6 origin of measurements. The remaining 9, 6 betas (3 for each group) and 3 alphas (for one group only), are to be estimated. Since there were 54 unique elements in the input data (each group had 15 correlations, 6 standard deviations and 6 means) and 27 parameters to be estimated, the model had 27 degrees of freedom (54 − 27 = 27). The chi square for M1 is 33.64 with 27 degrees of freedom, indicating that the fit is acceptable.

A more restricted model (M2) was then tested by fixing the parameter β_{64} to zero for the Regular group. M2 is in effect a quasi-simplex model for the Regular group. This new model yielded a chi square of 34.92 with 28 degrees of freedom. The difference between the chi-square values for M1 and M2 is not significant, indicating that the additional restriction imposed is reasonable, that is, that a quasi-simplex model is indeed adequate for the Regular group.

To determine whether the quasi-simplex model was adequate for both groups, the next model tested (M3) was one in which the beta parameter β_{64} was set to zero for the Immersion group as well. This model is also acceptable, as indicated by both the chi square and the chi-square difference (M3 − M2).

5.1 Invariance of the beta parameters

A restriction was then imposed on M3 by constraining the parameter β_{54} to be invariant across the two groups. The resulting model (M4) yielded a chi square of 34.93 with 30 degrees of freedom. The difference between the chi-square values for M4 and M3 with 1 degree of freedom is not significant, indicating that the hypothesis that the two groups have equal β_{54} parameters cannot be rejected. This establishes that the relationships between Grade 5 and Grade 4 are the same in both groups.

The next model tested (M5) is a test of the hypothesis that the relationship between Grade 5 and Grade 6 are the same for the two groups. This model is consistent with the data ($\chi^2 = 34.94$, d.f. = 31). Furthermore, the difference between the chi-square values for the last two models (M4 and M5) is only 0.01. Consequently, the two non-zero beta coefficients, β_{54} and β_{65}, can be regarded as equal over the two groups. The conclusion is that the entire structural relationship (except the intercepts) of the true variables is identical for both groups (see Table 3).

5.2 Differences between the alpha parameters

Once it has been shown that the betas are equal for the two groups, that is the use of a common regression coefficient is appropriate, it is meaningful

Mathematical Modelling of Education Data

Table 3. Chi square difference tests

Model	χ^2	d.f.	p	χ^2 d.f. = 1	p
(1) M1:	33.64	27	0.17	—	—
(2) M2: M1 and $\beta^R_{64} = 0$	34.92	28	0.17	1.28	n.s.
(3) M3: M2 and $\beta^I_{64} = 0$	34.93	29	0.21	0.01	n.s.
(4) M4: M3 and $\beta^R_{54} = \beta^I_{54}$	34.93	30	0.24	0.00	n.s.
(5) M5: M4 and $\beta^R_{65} = \beta^I_{65}$	34.94	31	0.29	0.01	n.s.
(6) M6: M5 and $\alpha_5 = 0$	42.11	32	0.11	7.17	0.01
(7) M7: M5 and $\alpha_6 = 0$	38.23	32	0.21	3.29	0.10

to consider the parameter alpha as an estimate of the difference between the two programmes.

A further model (M6) was then constructed to test the hypothesis that the difference between the two programmes is zero at Grade 5. This model yielded a chi square of 42.11 with 32 degrees of freedom, a substantially worse fit than M5. The difference between the chi-square values for M6 and M5 is 7.17 with 1 degree of freedom, which is significant at the 0.01 level. This would indicate that the difference between the two programmes at Grade 5 ($\alpha_5 = 2.52$) is statistically significant.

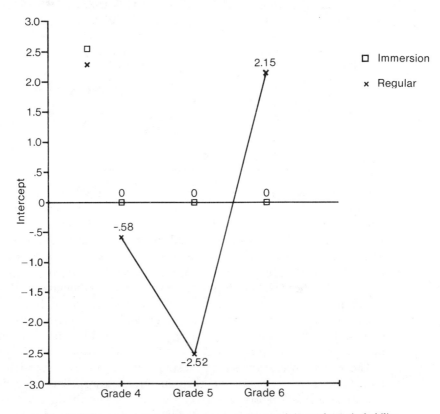

Fig. 2. Differences between the two groups in growth in mathematical ability.

The next model constructed (M7) tests the hypothesis that the difference between the two programmes at Grade 6 (α_6) is zero when the initial differences are controlled for. This model yielded a chi square of 38.23, with 32 degrees of freedom. However, the difference between the chi-square values for M5 and M7 is 3.29 with 1 degree of freedom ($p = 0.10$) which would indicate that the difference between the two programmes at Grade 6 ($\alpha_6 = 2.15$) is marginally significant (see Table 4).

Table 4. Maximum likelihood estimates and standard errors

Parameter	Model M5	
v_5	72.97(0.97)	
v_6	67.95(0.91)	
v_3	61.38(0.79)	
v_4	58.52(0.85)	
v_1	49.38(0.76)	
v_2	47.82(0.77)	
λ_6	0.85(0.06)	
λ_4	1.04(0.07)	
λ_2	0.96(0.07)	
β_{54}	1.20(0.08)	
β_{65}	0.93(0.07)	
ζ_6	13.53(5.68)	
ζ_5	10.46(3.34)	
σ_ζ^2	64.87(8.67)	
	Regular	Immersion
α_6	2.15(1.17)	0
α_5	−2.52(0.93)	0
κ_4	−0.58(1.36)	0
$\chi^2 = 34.94$	d.f. = 31	$p = 0.29$

To summarise, the analysis shows that the measurement properties of the tests were the same for the two groups. In both groups the relationships between Grades 4 and 5 and between Grades 5 and 6 from the aspect of mathematics ability can be described by the same quasi-simplex model. The results also indicate that the two groups started with equivalent mathematics ability. As shown in Fig. 2, the Immersion group had a significant higher level (2.52 points) of growth in mathematics ability than the Regular group at Grade 5; in the subsequent year, however, the Regular group exceeded the Immersion group by 2.15 points in its average growth. In comparison with the Regular group, the Immersion group grew more rapidly from Grade 4 to 5 and more slowly from Grade 5 to 6.

These results were arrived at by testing several models sequentially, each model being more restrictive than its predecessor. This procedure is essentially an exploratory one. To confirm the results obtained in this study it would be necessary to test the model on a different set of data.

REFERENCES

Bentler, P. M. (1982). Multivariate analysis with latent variables: causal modeling. In C. Fornell (ed.), *A Second Generation of Multivariate Analysis*. New York: Praeger.

Genessee, F. (1983). Immersion experiments in review. *Applied Psycholinguistics*, **4**, 1-46.

Hieronymus, A. N., King, E. M., Bourdon, J. W., Gossing, D., Grywinski, N. T. & Moss, G. L. (1974). *Canadian Tests of Basic Skills*. Toronto: Thomas Nelson & Sons (Canada).

Jöreskog, K. G. (1971). Simultaneous factor analysis in several populations. *Psychometrika*, **36**, 239-251.

Jöreskog, K. G. (1979). Statistical models and methods for analysis of longitudinal data. In J. Magidson (ed.), *Advances in Factor Analysis and Structural Equation Models*. Cambridge, Mass.: Abt Books.

Jöreskog, K. G. & Sörbom, D. (1981). *LISREL V—Analysis of Linear Structural Relationships by the Method of Maximum Likelihood*. Chicago: National Education Resources.

Sörbom, D. (1974). A general method for studying differences in factor means and factor structure between groups. *British Journal of Mathematical and Statistical Psychology*, **27**, 229-239.

Sörbom, D. (1978). An alternative to the methodology for analysis of covariance. *Psychometrika*, **43**, 381-396.

Sörbom, D. (1982). Structural equation models with structured means. In K. G. Jöreskog and H. Wold (eds), *Systems under Indirect Observation: Causality-Structure-Prediction*. Netherlands: North-Holland Publishing.

Swain, M. & Lapkin, S. (1981). *Bilingual Education in Ontario: A Decade of Research*. Toronto: The Ministry of Education, Ontario.

3
Modelling in Calculus Instruction—Empirical Research Towards an Appropriate Introduction of Concepts

Gabriele Kaiser-Messmer
Gesamthochschule Kassel (FRG)

SUMMARY

In the discussion about modelling in mathematics instruction it is often emphasised that modelling is not appropriate for the introduction of mathematical concepts. Rather one should use, as far as possible, concepts and techniques the students have been familiar with which for some years.

According to this point of view modelling should improve the students' ability to apply mathematical topics and techniques in real world situations, but not try to increase the students' mathematical knowledge. This point of view constructs a false contrast between the learning of mathematics and the learning of applying mathematics.

This chapter reports an empirical investigation that points out that there exists a relationship between the methodical procedure in the introduction of mathematical topics and techniques and the ability of the students to use these topics and techniques in the modelling process. The investigation uses the example of the concept of the derivative to examine the following questions.

(1) How far do students reach central skills which are to interpret mathematical concepts in the real world and to define real world situations with mathematical concepts by an application-oriented introduction of concepts in the modelling process?
(2) How far do students perform these real world interpretations or, accordingly, these mathematical descriptions after a mathematical introduction of these concepts as 'transfer'?

By this means, it is intended to compare some possibilities to introduce the concept of the derivative, for example rates of change, limit of the slopes of the secant line with regard to these questions.

This investigation allows us to draw the following conclusion: the development of abilities, which are central for the modelling process, strongly depends on an appropriate introduction of the mathematical topics and techniques and further modelling examples absolutely belong to such an adequate introduction.

1. INTRODUCTION

In the discussion about modelling in mathematics it is often emphasised that modelling is not appropriate for the introduction of mathematical concepts. Rather one should use as far as possible topics and techniques the students have been familiar with for some years. For example Burghes and Huntley formulate:

> We must equally warn you not to expect them (the students) to be capable of using newly assimilated mathematics in a modelling context. They will be happy and willing to use familiar mathematical concepts and topics, but will find it very difficult indeed to use a topic just covered ...—in modelling you should not be trying to increase mathematical knowledge, but to improve the ability to apply known mathematical topics in practical situations. In fact, whenever practical, we try to avoid any techniques learned during the previous year. (Burghes and Huntley, 1982, p. 739f)

Oke similarly writes:

> In an introductory course in modelling, the level of mathematics assumed should be several years below that of current achievement. (Oke, 1984, p. 91)

This point of view constructs a false contradiction between the learning of mathematics and the learning of applying mathematics. In what follows I will describe an empirical investigation which points out that there exists a relationship between the method of introducing mathematical concepts and the ability of the students to use these concepts in real world examples and the modelling process. The research was carried out using the example of the concept of the derivative. This investigation allows us to draw the following conclusion: the acquisition of abilities which are central for the modelling process strongly depends on an appropriate introduction of the mathematical concepts and techniques and further modelling examples absolutely belong to such an adequate introduction.

2. METHODOLOGICAL PROCEDURE IN THE EMPIRICAL INVESTIGATION

I used the case study method for the empirical research, as it is described for example by Stake (1978). Empirical-statistical methods were not used, because in the last few years the limitations of these methods have been pointed out (see e.g. Stenhouse, 1982).

It has been debated, for example, whether the results of empirical-statistical research are meaningful, since it is impossible to isolate and control all the relevant variables which influence complex situations like classroom instruction. Further it has been pointed out that actual theory deficits in the didactics of mathematics instruction do not allow research based on hypotheses which is implied by empirical-statistical methods.

Case studies aim at a description of the complex classroom situation in its entirety. They try to illustrate certain regularities within the exemplary description of individual cases in order to come to hypotheses. Therefore the limitations of the following investigation are evident: the investigation contains observations about individual cases and at first it is only possible to confirm facts about individual cases. But it is plausible to generalise the observations to a certain extent.

For the following investigation I have chosen the concept of the derivative, because it is on the one hand a central concept for mathematics instruction in the 'gymnasialen Oberstufe', which can be compared with the sixth form in the British school system. On the other hand, the usual methods of introduction are typical for different orientations of mathematics instruction, that is applications versus pure mathematics.

I used the following method in the research: I did not test the relationship between the introduction of concepts and the applying and modelling abilities with a modelling example, because the modelling process is influenced by a lot of uncontrollable factors like foreknowledge and previous experiences of the students, mathematical knowledge from other areas like calculus. Rather I carried out the research on component skills, which are relevant in particular phases of the modelling process, as follows.

(a) Within the phase of model formulation the ability of the students to define mathematically a real world situation with the difference quotient and the derivative.
(b) Within the phase of the interpretation of the results the ability of the students to interpret difference quotient and derivative in a real world situation.
(c) Within the phase of mathematical reasoning in the model the ability of the students to use the derivative to solve a real world problem.

The comprehensive research of Treilibs to a 'general modelling ability' and to 'component skills' in the phase of model formulation suggests that there exists a close correlation between the 'component skills' necessary in particular phases and a 'general modelling ability' (see Treilibs, 1979).

In nine classes of the 'gymnasialen Oberstufe' I carried out a test with approximately 150 17-year-old students (see Fig. 1), which should examine how far the students have the above-mentioned component skills. Thereby tasks 4 and 5 refer to the ability to mathematically define a real

TEST FOR THE DERIVATIVE

1. Which ideas do you connect with the concept of the difference quotient?
2. Which ideas do you connect with the concept of the derivative?
3. Interpret geometrically both concepts with your own mathematical example.
4. Interpret both concepts with your own real world example.
5. The atmospheric pressure decreases with the altitude. A certain atmospheric pressure belongs to each altitude.
   Define mathematically the concepts of 'mean incline of atmospheric pressure' and 'local incline of atmospheric pressure' with the difference quotient and the derivative.
6. Declare the real world meaning of x and $f(x)$

x	$f(x)$	$\dfrac{f(x+h)-f(x)}{h}$	$\lim\limits_{h \to 0} \dfrac{f(x+h)-f(x)}{h}$
		Average velocity of the growth of a bacteria culture	Instantaneous velocity of the growth of a bacteria culture

7. The amount of petrol, used by a car, depends on the travelled distance e.g. like the plotted graph.
   We denote the distance with s (in km) and the corresponding petrol consumption with $f(s)$ (in l).
   Interpret
   $$\dfrac{f(s+h)-f(s)}{h} \quad \text{and} \quad \lim_{h \to 0} \dfrac{f(s+h)-f(s)}{h}$$

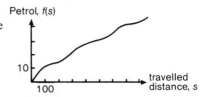

8. A cylindrical vessel contains particles of a not homogenous scattered liquid.
   We denote the volume of the vessel up to a (variable) position by x (in cm³), e.g.

   Further, we denote the (approximate) number of particles, which are in the volume x with $f(x)$ (the derivative exists). We examine a 'snapshot', that means we do not consider the time factor.

   Interpret
   $$\dfrac{f(x+h)-f(x)}{h} \quad \text{and} \quad \lim_{h \to 0} \dfrac{f(x+h)-f(x)}{h}$$

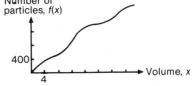

9. The production of goods needs certain efforts, which yield costs (e.g. wages of the employees, operating costs of the machines, rent for the buildings, cost for the raw materials). A certain sum of costs belongs to each produced amount of a certain commodity.
   The corresponding function linking a commodity with costs is called the 'cost function'. We denote the cost function with c. Then $c(x)$ = cost to produce x units for a certain commodity.
   Where is it profitable to raise production? Try to describe the area using the derivative (or slope of the function).

Fig. 1. Test.

world situation with a difference quotient and a derivative (aspect (a)), tasks 6 to 8 to the ability to interpret a difference quotient and a derivative in real world situations (aspect (b)), task 9 to the ability to solve an easy real world problem with the derivative (aspect (c)). Tasks 1 to 3 should demonstrate the students' conceptions of a difference quotient and a derivative, which might be helpful within the interpretation of the results to questions 4–9. Aspect (c) will not be considered, because the results to question 9 were not very meaningful.

All the students had learned the concept of the derivative 1–2 months earlier, but in very different ways.

In order to keep the description clearly arranged and to avoid distortion through strong differences in the ability level of the courses, I will restrict the presentation to four courses. These four courses were on a similar ability level, but the derivative had been introduced in different ways, as follows.

— Course A1: The difference quotient was introduced as rate of change in several real world examples, the concept of the derivative was developed as local rate of change within a continuous real world situation (average and instantaneous speed), the newly introduced concepts were interpreted in several real world situations.
— Course A2: Similar procedure as in course A1; but discrete situations were emphasised along with the difference quotient, the concept of the derivative was introduced in a different terminology and the newly introduced concepts were not interpreted in real world situations.
— Course G1: The difference quotient and the derivative were developed out of the local geometrical problem of the secant and the tangent line with detailed graphical illustrations.
— Course G2: Similar procedure as in course G1, but global conceptions were emphasised; it was started with graphical differentiation, the derivative was then developed along with the problem of the secant and the tangent line, immediately followed by detailed global examinations of functions.

In all four courses the derivative was defined as the limit of the difference quotient. In the following I will denote the courses A1 and A2 as 'applied courses' and the courses G1 and G2 as 'geometrically oriented courses'.

I calculated precise percentages considering the evaluation of the students' answers. However, these numbers indicate only rough tendencies and should not be overestimated, because the underlying population size was very small. Therefore I will additionally analyse the students' answers for underlying misconceptions or false patterns for the interpretation of the concepts. As the answers of the students were influenced by their previous experiences, I asked the students for the origin of their interpretations after the test.

3. RESULTS OF THE EMPIRICAL INVESTIGATION

The students' answers to the questions, which ideas they connected with the difference quotient and the derivative and the given graphical illustrations (questions 1–3) indicated that the different methods of introducing the difference quotient and the derivative led to very different conceptions of them by the students.

With the students of the applied course A1 no uniform conception of the difference quotient and the derivative could be recognised. The students had not only local geometrical ideas about the difference quotient (slope of the secant line), but also formal and applied conceptions (rate of change). With the derivative, ideas dominated referring to functions or calculations as opposed to local geometrical conceptions as the slope of the tangent line.

In the case of the applied course A2 the students' conceptions of the difference quotient fell into two groups, namely local geometrical concepts (slope of the secant line) and applied concepts (rate of change). As regards the derivative, all students had local geometrical ideas as the slope at a point.

In the case of the geometric courses G1 and G2, the differences were obvious, too: from the difference quotient many students of the course G1 had local geometrical conceptions as slope of the secant line or formal ideas, along with the derivative dominated local geometrical ideas as the slope at a point.

Compared with this, most students of the course G2 had only local geometrical ideas of the difference quotient. Referring to the derivative they had global ideas, that means they mainly saw the derivative as a function, as a means to examine functions and sketch graphs.

The geometrical illustrations of difference quotient and derivative (question 3) confirm the differences in the students' conceptions. These differences had severe consequences for the students' work in the applied tasks.

The students' abilities to define mathematically a real world situation with difference quotient and derivative were tested with questions 4 and 5. The following results were obtained in the more demanding task to construct a real world context on their own, in which difference quotient and derivative were relevant and to define mathematically this real world context with these two concepts.

Table 1. Answers to question 4

	A1	A2	G1	G2
Statement of a real world example	88%	57%	59%	19%
Correct precision of the real world example	67%	50%	30%	0%

Most students of the applied course A1 were able to specify a real world example and to define correctly these examples with the difference quotient and the derivative. Compared with this, the students of the geometrical course G2 almost completely failed. Little more than half of the students of courses A2 and G1 were able to assign real world examples.

The results of the easier task to define mathematically a given simplified real world situation differed widely, too.

Table 2. Answers to question 5

	A1	A2	G1	G2
Correct description (formal)	82%	0%	59%	13%
Correct description (only verbal)	6%	67%	6%	6%
False description	6%	20%	35%	31%
No answer	6%	13%	0%	50%

Nearly all students of the course A1 mathematically defined the situation in a correct manner, just as two-thirds of the students of the courses A2 and G1. Only one-fifth of the students of course G2 were able to do this.

It is possible to recognise misconceptions in the students' answers: some typical examples were as follows.

The students of the course A2 connected with the derivative a general formula for the calculation of the slope, a certain kind of mean calculation, wherefore they defined the mean incline of atmospheric pressure with the derivative. For them the difference quotient made specific statements about a piece of the function, therefore they defined the local incline of atmospheric pressure with the difference quotient.

With the help of the difference quotient one calculates the slope of a function within a certain straight line, y_0 to y_1. The derivative specifies the slope of the function in its entirety. As for example the quadratic function hasn't the same slope at each point of the function, the slope, which is assigned by the derivative, is denoted with mean incline of atmospheric pressure ... The difference quotient refers to a certain piece of the function and therefore it is more precisely aligned with this part of the function; accordingly it describes the local incline of atmospheric pressure. (Rough translation.)

Many students of the course G1 could not imagine that it was possible to define mathematically situations like incline of atmospheric pressure. Their answers seemed to be guessed from the description of the situation.

Local incline of atmospheric pressure is the slope at a certain point of the function. Mean incline of atmospheric pressure is the derivative, which describes the slope at each point of the function. (Rough translation.)

The students of the course G2 had no idea how to describe situations like this in a mathematically precise way. The following statement shows this.

Mean incline of atmospheric pressure is the function, local incline of atmospheric pressure is the first derivative.

The students' abilities to interpret the difference quotient and the derivative in a real world situation were tested with questions 6–8. The results of the question which referred to an everyday situation, and the question which dealt with a scientific situation, differed widely. First to the results of the interpretation of the difference quotient and the derivative in an everyday situation.

Table 3. Answers to question 7

	A1	A2	G1	G2
Both interpretations correct	50%	40%	59%	13%
Only interpretations of difference quotient correct	38%	27%	30%	0%
False interpretations	6%	14%	12%	44%
No answer	6%	20%	0%	44%

Hardly any differences between the courses A1, A2 and G1 existed, only the students of the course G2 again failed completely. The answers globally show that many students of the courses A1, A2 and G1 were able to interpret the difference quotient and the derivative in an everyday situation. However, a detailed analysis showed that many students had difficulty only with the interpretation of the derivative, not with the interpretation of the difference quotient. In the following I will illustrate some problems with typical answers of the students. For example, some students of the courses A1 and G1 had problems with the interpretation of the 'h', which becomes smaller and smaller. A typical statement of a student from course A1:

> Difference quotient: 'average petrol consumption'; derivative: 'average petrol consumption referring to a certain distance'.

The difficulty of the students of the course A2 was on a completely different level, they were strongly influenced by the terminology used in the previous instruction. So, many students connected a general formula about petrol consumption on the whole distance with the difference quotient, whereas the derivative not only describes the petrol consumption at a certain point but also the general petrol consumption.

> Difference quotient: 'Here the petrol spent $f(s)$ per distance s is given in general'; derivative: 'Here a description of an approaching point (limits) is carried out, made at an optional point. I can not only determine the general petrol consumption but also the petrol consumption at a certain point.' (Rough translation.)

In the ensuing discussion some students mentioned that they had remembered the calculation of the petrol consumption, common in everyday life.

In the case of the interpretation of the difference quotient and the derivative in a scientific situation, in which the students were lacking such previous experiences, the results of all courses became distinctly worse.

Table 4. Answers to question 8

	A1	A2	G1	G2
Correct interpretations	31%	0%	6%	0%
Partially correct interpretations	19%	33%	18%	6%
False interpretations	44%	27%	70%	19%
No answer	6%	40%	6%	75%

With one slight exception only, students of the course A1 gave correct interpretations of the difference quotient and the derivative as 'average density of the particles' and 'local density of the particles', that is about a third of the course. Half of the answers were false, but it was not possible to recognise prevailing misconceptions. Nevertheless a third of the students of the course A2 gave partially correct interpretations. Again the students' problems were caused by the terminology used in previous instruction, which led to a confusion about general, global versus specific, local statements. One typical answer:

> Difference quotient: 'Here a concrete volume is calculated that is in principle only the general value of the volume, that means we calculate an average value...', derivative: 'Here we carry out a description of changes with the help of a complete description of all relations in order to come to assertions about a general relationship between x and the number of the particles, that means we take a general value and carry out a discussion of the limit of the function'. (Rough interpretation.)

In the course G1 more than two-thirds of the answers were false, for example some students gave interpretations related to the time:

> Difference quotient: 'Average number of particles per volume'; derivative: 'number of particles per volume at a certain moment'.

Other statements of students of the course G1 only showed that the students simply did not know how to interpret such situations, because of lacking experiences from the preceding instruction or other teaching subjects, for example: Difference quotient: 'volume of the vessel'; derivative: 'number of particles'.

The results of another task (question 6) point to great differences among the different courses in the students' abilities to apply mathematics. In this task the students should reconstruct the real world meaning of the underlying function from the given meaning of the difference quotient and the derivative, as it were the inversion of the previous two tasks. The following results were attained.

Table 5. Answers to question 6

	A1	A2	G1	G2
Correct description	82%	40%	24%	6%
False description	18%	47%	71%	57%
No answer	0·	13%	6%	38%

Nearly all students of the course A1 were able to specify the real world meaning of x and $f(x)$. In the other applied course A2 nearly half of the students succeeded in giving the correct answer. In the geometrically oriented course G1 only a quarter of the students gave correct descriptions, the course G2 failed again.

4. INTERPRETATION OF THE RESULTS

In my opinion the result of the research allow the following theses.

Thesis 1

It is possible extensively to reach the following abilities, central for modelling in calculus instruction, along with an applied introduction of concepts using adequate and favourable methods:

— the ability to define mathematically a real world situation with difference quotient and derivative
— the ability to interpret the difference quotient and the derivative in a real world situation.

In my opinion the results of the applied course A1 make it possible to establish this thesis. The results of the other applied course A2 are worse throughout. This shows that the kind of methodical procedure, the terminology used play an important part. I will refer to this in thesis 3.

Thesis 2

In the case of a pure mathematical introduction of difference quotient and derivative the students have great difficulty using these concepts in real world situations, not only to define mathematically real world situations, but also to interpret these concepts in real world situations. A considerable group of students is able to apply these concepts in real world situations as 'transfer'. However, they succeed to a smaller extent than the students who had learned the concept of the difference quotient and the derivative application-oriented.

This thesis is based on a comparison of the results in tasks 4–8 of the applied course A1 with the results of the geometrically oriented courses G1 and G2. The course A1 reached altogether better results than the two other courses, whereby the differences between the two geometrically oriented courses G1 and G2 were considerable.

Thesis 3

The ability to define mathematically a real world situation with the difference quotient and the derivative and to interpret these concepts in real world situations, which are central for modelling in calculus instruction, are influenced not only by the method of introducing these concepts—that means applied versus mathematical—but also by the method.

— In connection with an applied introduction of difference quotient and derivative, obstacles exist for the development of such abilities, such as extreme differences between an applied introduction and the mathematical continuation, a missing phase of real world interpretations after the introduction of the concepts, or a strong emphasis on discrete processes of change in the introduction phase of the concepts.

— Along with the mathematical introduction of the difference quotient and the derivative, methods exist which hinder real world interpretations and methods, that do not obstruct such interpretations. Local geometrical conceptions of the difference quotient and the derivative as slope of the secant and the tangent line do not promote such real world interpretations without extra effort, but they do not obstruct such interpretations.

Compared with this a strong emphasis on the examination of functions and on global aspects, occurring very often in connection with the introduction of the derivative over graphical differentiations, leads many students to conceptions which obstruct such real world interpretations.

The first part of the thesis is based on the difficulty of the applied course A2 with the tasks to define mathematically a real world situation and to interpret the difference quotient and the derivative in a real world context. These problems were really significant and more serious than the problems of the applied course A1. Especially the terminology used of the zero sequence in the introduction of the derivative, strictly separated from the concept of the rate of change dominating in the introduction phase, led to a confusion of local and global aspects and further to a confusion of average and local rates of change.

The second part of the thesis relies on the results of the course G1, which were not bad. The students of this course had local geometrical conceptions of the difference quotient and the derivative, which obviously did not obstruct interpretations of the difference quotient and the derivative as average and local rate of change. The global conceptions of the students of the course G2 from the derivative as a function, as a means to examine functions and sketch graphs, turned out to be a serious obstacle for local real world applications of the difference quotient and the derivative.

These results point to the central role of the method of introducing mathematical concepts. It strongly influences which mathematical concepts and methods can be used in modelling examples. The introduction of mathematical concepts and methods in real world situations and the practice of their use in a real world context can definitely promote and facilitate their application in modelling examples. Treilibs drew similar conclusions of his empirical research and as early as 1980 formulated these at ICME 4 in Berkeley:

> Explicit steps must be taken to develop modelling ability; it is inefficient to merely present large numbers of standard models and hope that the students will learn modelling skills 'by example'. Such steps should

include the practice of the component skills of modelling ..., and the development of an awareness of the need for, and nature of, validation procedures. (Treilibs, 1983, p. 298)

In my opinion the introduction of concepts can already promote modelling abilities, but an akward, not appropriate, introduction can obstruct such abilities. Again these results refer to the central role of the teacher, who can definitely influence the development of modelling abilities with an adequate methodical procedure.

REFERENCES

Burghes, D. N. & Huntley, I. (1982). Teaching mathematical modelling—reflections and advice, *International Journal of Mathematical Education in Science and Technology*, **13** 735–754.

Oke, K. H. (1984). Mathematical modelling—a major component in an MSc course in mathematical education. In J. S. Berry *et al.* (ed.), *Teaching and Applying Mathematical Modelling*. Ellis Horwood, 86–95.

Stake, R. E. (1978). The case study method in social inquiry, *Educational Researcher*, **7** 5–8.

Stenhouse, L. (1982). Pädagogische Fallstudien: Methodische Traditionen und Untersuchungsalltag. In D. Fischer (ed.), *Fallstudien in der Pädagogik*. Faude, 24–61.

Treilibs, V. (1979). *Formulation Processes in Mathematical Modelling*, Shell Centre for Mathematical Education.

Treilibs, V. (1983). The mathematical modelling abilities of sixth form students. In M. Zweng *et al.* (ed.), *Proceedings of the Fourth International Congress on Mathematical Education*. Birkhäuser, 297–298.

4

Assessment in Mathematical Modelling

K. H. Oke
Polytechnic of the South Bank, UK

and

A. C. Bajpai
Loughborough University of Technology, UK

SUMMARY

This chapter discusses *assessment criteria* and *methods* that have evolved from research in modelling processes and other considerations based on the teaching experiences of the authors. The two main forms of assessment which are considered are *written examinations* (formal fixed-time) and *course-work assignments* (taking from 12 to 50 hours to complete). In general agreement with others, written examinations are considered to be a most inappropriate method of assessment; however, with large classes or for other reasons, this method has been used and the advantages/disadvantages are discussed with illustrative examples. Course-work assignments, on the other hand, are considered to be a most appropriate form of assessment and an indication of the standards reached by teachers in an MSc course in Mathematical Education as well as some undergraduate work is provided.

Criteria for guiding assessment, no matter what the assessment mode, are examined. Briefly, the criteria amount to suggestions for credit to be given for the following: initial interpretation of problem, generation of relationships consistent with initial objectives, technical competence in mathematics, rational simplifications based on assumptions, recognition of a solution, conclusions, and overall presentation. The way in which such a set of criteria is consistent with current understanding of modelling processes is also examined.

Finally, the arguments for and against *formal marking* rather than *impression marking* are studied and illustrated in a selection of assessments. It is shown that, in common with the findings of others, formal markers often adjust their marks under some headings in order to compensate for their overall impression of a piece of work.

1. INTRODUCTION

Most authors agree that a *formal written examination* is the most inappropriate method of assessment in mathematical modelling. Occasionally it has been used, see for instance the comments of Burley and Trowbridge (1984), in view of the large number (fifty or sixty) of students involved. With large numbers of students, staff resources usually do not stretch to the much more time-consuming process of reading and marking the more appropriate project type of assignment (*course-work*). Modelling has also been assessed by written examination in the second level course, TM 281 'Modelling by Mathematics' which was introduced by the Open University in 1977. When the MSc in Mathematical Education (CNAA) was introduced at the Polytechnic of the South Bank, assessment was by written examination and by course-work (Oke, 1980; 1984a). In the latter case, the reason for the inclusion of a written examination paper was to attempt to balance the assessment modes in what was a completely new experience (running an MSc Math. Ed.) both for South Bank and for the CNAA.

The marking of either written examination papers or of course-work type assignments is difficult because of the uncertainty of what criteria to use. Such criteria depend largely on one's understanding of the modelling process and on what students find most difficult in this process. Several authors, for example Treilibs (1979), McLone (1979), Burkhardt (1979, 1981), James & Wilson (1983), Berry & Le Masurier (1984), and others, have reported on what appears to be a common set of students' difficulties, namely:

> General lack of confidence.
> Loathness to simplify.
> Lack of skills in approximating and estimating.
> Inability to generate mathematical relationships.
> Knowing when to stop.
> Weakness in report writing.

Berry and O'Shea (1982) report on their experiences of assessing the mathematical modelling project set in the Open University's MST 204 unit. This unit was presented to students for the first time in 1982, and it is the results for that year which are analysed by Berry & O'Shea, with further discussion presented by Berry & Le Masurier (1984). A *formal marking scheme* is used, and out of a possible total of 100, 35 possible marks are awarded to 'initial model' and 'formulation' (taken as a combination). Relatively few marks (10) are given to 'data', since the experiences of the OU with other projects has shown 'that students in difficulty may attempt to accumulate marks by amassing vast amounts of data'. A relatively high mark (20) is given to 'revisions to the model', thus encouraging students to be critical of their first attempts and to make some improvements. Berry & O'Shea report favourably on the consistency of project markers, and most interestingly Berry & Le Masurier report that *marks formed by overall impression* were very close most of the time to those marks obtained by following the formal marking scheme.

The articles by Berry & O'Shea (1982) and by Berry & Le Masurier (1984) are amongst the most detailed recently published on assessment procedures in modelling. Their marking scheme represents an *additive* one, like most marking schemes, but Hall (1984) suggests that a product scheme of marking might be more realistic and also provide a more uniform method of marking different individual projects (course-works). Hall argues that no credit should be given if a vital component of modelling ('content', 'presentation', or 'drive') is absent or is very badly done. So, Hall is definitely recommending strict standards according to a *formal marking scheme*. He further recommends that 'double-blind' marking is used (two markers neither of whom has seen the other's marks, mark each project). Several authors would seriously question the advocacy of formal marking, and would rather argue for *impression marking*; for example, Burghes & Huntley (1982) recommend 'marking by interview' where groups of students discuss their course-work report with a lecturer and a mark is *jointly agreed*. As mentioned earlier, Berry & Le Masurier found that impression marking has led to close agreement with marks awarded according to a formal marking scheme.

Subsequent sections in this chapter discuss the points mentioned above in more detail and also show how the work on *formulation–solution processes* (in the next chapter) has an important bearing on assessment in mathematical modelling.

2. IMPLICATIONS OF FORMULATION–SOLUTION PROCESSES FOR ASSESSMENT

Associated with any assessment form are the issues of *formal* and *informal grading* (the latter is sometimes referred to as impression marking). As discussed in Section 1, there are arguments for and against each method of grading. These issues are taken up again in the subsequent sections of this chapter, but suffice it to say at this juncture that although there are strong arguments in favour of informal grading (even for externally assessed assignments), a formal marking scheme which awards marks for each of well-defined attributes or sections of a student's modelling attempt may be commended for the lecturer inexperienced in the teaching and assessment of mathematical modelling.

Some key considerations which guide assessment, no matter in which form or whether a formal marking scheme is used, are indicated by the findings of the next chapter on formulation–solution processes.

As pointed out in the next chapter, it is the *relationship level graph* (RLG) rather than the *concept matrix* (CM) that has provided the deeper insights into modelling processes. Consequently, the results of analysing formulation–solution processes using RLG are the most relevant in providing guidance for assessment. The RLG has shown that formulation and solution are intimately interwoven (carrying out some mathematics prompts the need for further understanding of the problem—generation of further level 0 relationships). So, formulation and solution may best be marked together. Analysis, using RLG, of students' attempts at modelling has shown that although 'interpretation' and 'validation' are often an integral part of 'formulation–solution', they can be more naturally separated out for marking. The RLG has also shown, through demonstrating relationship generation and the possible evolution of sub-problems, that model development and improvement take place naturally; consequently, it is unreasonable to insist on students *in all cases* to make a separate development of models in a hierarchical sense. Both the CM and RLG show that simplifying assumptions, relationships, variables and constants are generated naturally with the development of a model(s), and so it is artificial to ask for a list of such items in the initial part of a report—such items could only be listed with *hindsight* and *out of their natural context*. The latter point is not encouraging lack of clarity; on the contrary, students should be encouraged to identify most clearly any assumptions and variables they create as they develop their model(s).

The above points may be summarised as follows.

(1) Formulation and solution are intimately interwoven, even in 'polished' model developments, and so are best treated as a single entity.
(2) Interpretation and validation can be more easily separated out for marking. A warning must be issued even here, though, since these latter activities are a vital part of the modelling process and are themselves often integrated with formulation–solution activities.
(3) Improvement of the model can take place in natural development and so it is unreasonable to insist on separate treatment.
(4) Sub-problems are often only identified with hindsight, consequently it is unreasonable to ask for separate treatment of each.
(5) Simplifying assumptions, relationships, variables and constants are generated naturally with model development. Consequently it is artificial to ask for a list of such items at the outset.

Additional considerations based on the authors' experience in assessment which incorporate points (1)–(5) above are the following.
Credit to be given for:

(A) Interpretation of problem, including clear statements of initial objectives
(B) Generation of relationships consistent with inital objectives
(C) Technical competence in mathematics in generating additional relationships
(D) Rational simplifications making clear any assumptions made
(E) Recognition of a solution—ability to interpret and validate. Checking for logical errors.
(F) Conclusions and general discussion—awareness of strengths and weaknesses of model development, suggestions for further work

(G) Overall presentation—ability to communicate clearly in written form; clear diagrams and sketches

In the subsequent sections the fundamental points made earlier will be embodied in discussions on assessment of examination papers and of course-work assignments. Additional considerations specific to a group of students as well as the form of assessment will also be identified.

3. WRITTEN EXAMINATIONS

This section refers to written examinations in mathematical modelling and, in particular, illustrates with examples of questions set in the MSc Math. Ed. final year (second year) assessment.

The MSc Math. Ed., the only course of its kind in the public sector of higher education, started running in 1977. This part-time course is intended mainly for secondary school teachers and college of further education lecturers who have a degree or equivalent qualification in mathematics. Details may be found in Oke (1980, 1984a).

The examination paper, which is taken at the end of year 2, is of three hours duration. The paper consists of two sections. Three questions are to be attempted (1 hour per question), with *one* question only selected from Section A.

Section A (Seen one week before examination)
Three questions, each stating a practical problem, to be modelled from scratch. Only initial approaches are expected, but they must include some mathematics and interpretation. One question is based on a problem in the social and organisational area, one on physics/engineering area, and one on life sciences/biology.

Example (physics/engineering area)
Modern office blocks, particularly of the high rise type, have large glazed areas on the outside to permit entry of as much natural light as possible. By concentrating on the forces involved on an individual glass unit or pane, try to identify some key design features. Is there an optimum pane size, and if so, does double glazing affect this? In your development, consider simple models and make clear any assumptions you feel are necessary.

(June 1983 paper)

Section B (Unseen)
Approximately five or six questions, each based on general modelling and/or pedagogic issues. Essay type answers expected.

Example
Make out a case for teaching mathematical modelling, indicating clearly the level and background of the students involved. Refer to relevant articles as far as possible.

(June 1983 paper)

In order to provide an indication of the extent of the initial modelling development that is expected in response to a Section A type question, the following outlines a possible approach to the office block glazing problem above:

Office block glazing (Section A, June 1983)
Outline notes on possible approach.

Consider single-glazing. Size of glass-pane is limited by risk of glass breakage; pane needs to be as large as possible to allow maximum amount of light entry—too many panes over a large area will involve loss of light entry due to area of supporting frames. Consequently, there appears to be an optimum size for a given pane.
   Key methods by which pane is assumed to break:

(a) Wind causing flexure.
(b) By crushing under own weight.
(c) Thermal cracking—pane not allowed to expand (or contract) in frame.

With a well-designed frame, it can be assumed that (c) will not occur. Before (b) takes place, whole side of high-rise office block would consist of single pane of glass! Wind forces causing flexure, as in (a) seem to be the single most important cause of breakage (ignoring accidents).
   Assuming frame is rigid on all four sides of pane, then problem reduces to 2-D stress type (assuming small displacements). If wind speed is v, $d/dt(mv) = \dot{m}v = (\rho A v)v = \rho A v^2$ can be assumed from Newton's second law to be force (normal on pane of area A). ($\dot{m} = \rho A v$ is flow-rate of wind, ρ is density of air). This approach would provide simplified boundary conditions. By solving the biharmonic stress equation, maximum stresses can be found (near centre of pane). The design would involve knowledge of maximum possible wind speed v (over the year, in a given location), so that maximum stress is much less (50% less?) than breaking (yield?) stress of glass involved. Hence size of pane. For double glazing, air is trapped between two panes of glass and would be partly compressed—this might

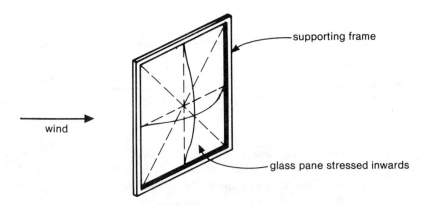

Fig. 1. Flexure due to wind forces.

strengthen structure and hence permit a larger unit for given wind speed; stressing of inner pane would also have to be taken into account.

So far, the mathematics that would be involved would be fairly complicated and beyond expectation in the time allowed (one week to prepare modelling approach, and one hour in the examination in which to write out the development). So, it is wise to consider an even cruder approach in order to get some upper-bound for stress at the middle of the pane.

Crude model

Consider a single-pane of glass, rigidly supported along upper and lower edges only, then problem reduces to one in 1D (see Fig. 2).

Maximum stress (at mid-point) would be greater than for 2-D model and hence would be an upper-bound. Solution follows from elementary beam theory, using resultant of air force and weight for external loading. Full credit would be given for a comparable development.

Note. If an approach along the lines of the above development were followed, then some attempt at solving the beam problem identified above would be expected.

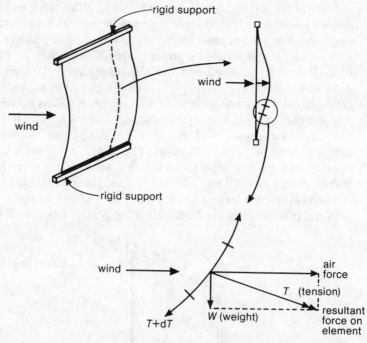

Fig. 2. Crude model.

Section A (one question) and Section B (two questions) are allocated equal marks by *informal (impression) marking*. It was decided by the marking team at South Bank that formal marking was inappropriate in view of the possibly wide variation in approaches that could be adopted in tackling any one question. For example, the outline approach provided above (Oke's) represents mainly initial formulation, with reasons, of a crude model; little mathematics is used or intended (elementary beam theory and solution of a differential equation is the most expected). Consequently most credit, for a comparable development, would have to be given to initial arguments of the type used above in creating a specific problem to be solved. On the other hand, a student may decide (this actually happened in one case in 1983) that only the briefest (half-page) discussion would suffice, and then proceed with a solution with some numerical values (from a textbook) being inserted. Credit would, in the latter case, concentrate on solution and interpretation. As a measure of the standards set for the course, the approach which has been outlined together with some solution and interpretation of the elemental beam would attract full marks; without the latter solution, a mark of two-thirds of the total would be awarded. Section B questions are marked as essays, where content, presentation, relevance and clarity in communication are given credit.

Clearly, it is not reasonable to expect an extensive modelling development for a Section A question. In fact, all that is insisted upon are points A, B, D and G with some attention paid to the remaining from the credit list provided in Section 2.

Examination papers for the years 1979–1980 had the same structure except that Section A was unseen. The poor standards achieved in Section A persuaded the teaching team to adopt a 'seen' approach from 1981 onwards, which resulted in considerable improvements in student performance. However, in view of the realistic expectation of few modelling activities being carried out in the time available and under the stress conditions of a formal fixed-time examination, it has now been decided to discontinue this mode of assessment from 1985 onwards. The reason for the inclusion of a written examination paper in the first place was an attempt to balance the assessment modes in what was a completely new experience (running an MSc Math. Ed.) both for the South Bank Polytechnic and for the CNAA.

Mathematical modelling is also assessed by course-work on the MSc and this mode will be the main mode of assessment in 1985 and subsequent years. The next section discusses course-work assignments, with illustrations of the assignments involved with the MSc and BSc Applied Physics courses at South Bank.

4. COURSE-WORK ASSIGNMENTS

4.1 M.Sc Math. Ed.

In the case of MSc Math. Ed., one course-work assignment is set towards the end of year 1. Originally, two assignments were set, but largely due to a policy of reducing the overall number of assessments on the course in all subjects, a concession had to be made in mathematical modelling.

This assignment consists of each student (teacher) finding their own problem, in any area they wish, and developing a mathematical model relating to this problem. Teachers are expected to define the learning aims appropriate to a level of student with which they are familiar, and to provide self-assessment questions for their students—these questions may test understanding of the developed model as well as test ability to extend or model a similar situation. Originally this course-work was assessed according to the following formal marking scheme.

(1) Statement of problem, to include how the problem was identified in the first place. — 10
(2) Learning aims (broad and specific teaching aims, including level of students for whom material might be wholly or partially appropriate) — 10
(3) Construction of model
(4) Analysis of model (including validation) } 50
(5) Discussion (general and conclusions) — 20
(6) Self-assessment question(s) (for intended students) — 10

Total 100

Assessments (1), (3), (4) and (5) would be appropriate to any modelling exercise, whereas (2) and (6) are specifically relevant to the teachers on the MSc course. Note that whether formal marking is used or informal (impression) marking is used, the above serves as a useful check-list. Note also, that in view of the comments made in Section 2, a further break-down of modelling activities is avoided although points A–G do provide an additional overall guide. As the teaching team gained experience in marking course-works, impression marking has taken over. This approach is further supported since teachers have considerable choice in how they present their work, and because of the completely free choice they have in the problem (which they find) to model.

The whole matter of assessment, regarding both examination papers and course-work assignments, has been discussed at length on the Advisory Committee for Mathematical Education (South Bank), chaired by Professor A. C. Bajpai. The committee agreed that mathematical modelling would be more appropriately assessed by course-work rather than by formal fixed-time examination. The external examiners of the MSc course have agreed that whilst a formal marking scheme for course-work can be of value, the most important criterion for judging a particular piece of work is based on knowledge of *standards* that have been developed as a result of running the course over several years. These 'standards' are established by 'impression' marking whereby the internal examiner, in final concurrence with the external examiner(s), arrives at a final mark (grade) by appraising the overall quality of a piece of course-work using points (A)–(G) as guidance.

Teachers are asked to find their own problem and to develop a modelling approach comparable in extent to some samples provided in the earlier part of the course. In other words, although a thoroughly competent development is expected, any attempts at elaborate mathematics and/or attempts at introducing an abundance of detail into an analysis is discouraged. Credit is given for a development that is consistent with the learning aims that must be identified at the beginning of each report. On the whole, teachers produce work within the reasonable perspectives outlined here. However, there are one or two exceptions where quite voluminous and over-ambitious reports have been presented; in the latter cases, excessive enthusiasm had led to attempts to study a problem in a manner which is much more appropriate to a team of professional modellers with much more time available. In the other extreme, some reports contain a large amount of descriptive material with little mathematical content, and consequently the benefits of modelling are barely achieved.

It is very important for students in their development of mathematical modelling skills to receive comments on their assessed work in order that they may improve on their weaknesses. A balance between encouragement and criticism is required, especially with part-time students where there is inevitably less contact between lecturing staff and students (teachers) than is the case with full-time students.

4.2 BSc Applied Physics

In contrast to the extensive course-work that is expected of the MSc Math. Ed. teachers, taking an average of 52 hours and where a problem has first to be found, course-work on mathematical modelling takes approximately 12–15 hours in the BSc Applied Physics. A problem, or set of problems, is presented to the physicists in the form of a problem statement.

Mathematical modelling was first introduced on the BSc Appd Physics degree four years ago. At present it is taken only in the second year of the course, but it is planned to include modelling in the first year as well from 1985 onwards. The subject forms a compulsory part of the curriculum and it is assessed; marks contribute towards the final part I of the degree.

The course-work assignment consists of a practical problem that is presented to the class which is then split into groups; the groups then work for two weeks (3 hours per week) as part of their normal course where contact may be made with a lecturer. At the end of the two-week period, students have an additional week in which to write up *group reports* in their own time.

In order to illustrate the assessment of this type of assignment, the groups referred to above who worked on the record player problem (see Oke (1981) for further details and also Sections 3 and 4 of the next chapter) will now be considered. A formal marking scheme was adhered to on this occasion as follows.

Group report to be in following format

(1) Problem statement
(2) Report on class discussion (initial clarifications of problem with a lecturer).

(3) Log consisting of minute by minute group development of model(s). This must be an honest and accurate record of what actually happened
(4) Report consisting of model(s) with interpretation of results based on (3) above
(5) Conclusions
(6) References if any

Marks awarded as follows:

	%
Overall presentation	20
Log (Section 3 above)	30
Main report (Section 4 above)	40
Conclusions (Section 5 above)	10
Total	100

The decision to assess each group, rather than individuals, seemed to be a natural one since groups worked together as teams. The disadvantage of assessing in this manner, however, is that the less able or less hard working get the same credit as the stronger members of their group. Little discord was observed on the latter point, although each group did tend to produce a leader. Most reports show evidence of a genuinely co-operative effort, at least to the extent of sharing the writing of sections amongst group members.

It was decided to assess according to a formal marking scheme by *triple-blind marking*; one marker was one of the authors (Oke), another was a moderately experienced lecturer in modelling (and its assessment), and the third marker was relatively inexperienced in modelling. The final mark awarded was an average of the three markers. The three markers independently observed the groups working in class time and made notes. The marks produced are shown in Table 1. Also shown in Table 1 is the maximum relative discrepancy (MRD) between markers, where

MRD = numerical value of maximum difference between markers
 ÷ average mark

(For example, marks for presentation for group 1 are respectively 13, 15, 14. Hence MRD = 2/14 = 0.4 (approx.)).

The table shows no consistent difference between the total marks given in any group across the groups, in fact there is surprisingly close agreement. However, there are more significant (although still not consistent) differences in the marks given to each section as shown by the higher MRD values. The most striking differences occur for marks awarded to the conclusions section; these differences (highest MRD is for group 3) will not contribute much to the total marks, however, since this section can at most contribute 10 out of 100 in weighting. No doubt the overall close agreement between the markers can be explained by the fact that all three were closely involved with the observation of the groups.

Table 1. BSc 2 Applied Physics course-work group marks: minimisation of sound distortion in a record player

	Group 1	Group 2	Group 3	Group 4
Presentation (max. 20)	13 15 (0.14) 14	11 13 (0.17) 12	11 16 (0.35) 16	16 12 (0.28) 15
Log (max. 30)	20 18 (0.11) 19	16 18 (0.17) 19	19 20 (0.05) 20	23 21 (0.09) 21
Report (max. 40)	27 22 (0.20) 27	16 18 (0.12) 18	24 24 (0.08) 26	35 35 (0.15) 30
Conclusions (max. 10)	5 8 (0.45) 7	4 5 (0.40) 6	6 3 (0.60) 6	3 4 (0.50) 5
Total	65 63 (0.06) 67	47 54 (0.15) 55	60 63 (0.13) 68	77 72 (0.08) 71
Average total	65%	52%	64%	74%

First number in each box: Oke's mark.
Second number in each box: moderately experienced marker.
Third number in each box: relatively inexperienced marker.
Number in brackets in each box: maximum relative discrepancy between markers.

Note that more pronounced differences in marking might have been predicted in view of there being no break-down in marks for the main report section, where the model(s) development takes place. That such close agreement amongst the markers (highest MRD is 0.20 for group 1) has been achieved is another instance of support for informal (impression) marking.

5. CONCLUSIONS

This chapter covers general points for guidance in the assessment of mathematical modelling assignments. The two main forms of assignment considered are written examinations and course-work. Illustrations of the points have been made by referring to the assessment methods used in the MSc Math. Ed. and BSc Appd Physics courses offered at the South Bank Polytechnic.

The overall implications of the chapter on formulation–solution processes for assessment as well as the presentation of a credit guidance list are covered in Section 2.

A subset of modelling activities is all that can be expected in a formal written examination and consequently *this form of assessment is not recommended*. The limited scope for assessing modelling in this manner is illustrated in the case of the MSc Math. Ed. in Section 3.

By contrast, the less stressful mode of course-work, where much more time is made available, is considered to be a *most appropriate form for assessment*. Examples of marking schemes used in assessing modelling assignments in the MSc Math. Ed. and BSc Appd Physics courses are provided in Section 4. Irrespective of the marking schemes considered, all points in Section 2 are expected to be covered for full credit to be given. A case for informal (impression) marking is made, where the assessor has an eye for attributes in the credit list appearing in some form or other in a course-work report. Formal marking schemes may best be used by inexperienced lecturers, although even then a large element of judgement is needed in attributing marks to any section. Close agreement is often achieved between several markers, even where a vaguely defined section is part of the marking scheme; this is illustrated in Section 4.2 in the marking of the record player problem. Such close agreement may well be due to lecturers (markers) being closely involved in observing students modelling a particular problem or may be due to lecturers working closely together as a team over several years (as in the case with the MSc Math. Ed.).

REFERENCES

Berry, J. & Le Masurier, D. (1984). OU Students Do It By Themselves. Proceedings of the International Conference on the Teaching of Mathematical Modelling, held at the University of Exeter. In: J. S. Berry, D. N. Burghes, I. D. Huntley, D. J. G. James & A. O. Moscardini (eds), *Teaching and Applying Mathematical Modelling*, Ellis Horwood, pp. 48–85.

Berry, J. & O'Shea, T. (1982). *Int. J. Math. Educ. Sci. Technol.,* **13**, 6, 715–724.

Burghes, D. & Huntley, I. (1982). *Int. J. Math. Educ. Sci. Technol,* **13**, 6, 735–754.

Burkhardt, H. (1979). *Bull. IMA,* **15**, 10, 238–243.

Burkhardt, H. (1981). *The Real World and Mathematics*, Blackie.

Burley, D. M. & Trowbridge, E. A. (1984). Experiences of Mathematical Modelling at Sheffield University. Proceedings of the International Conference on the Teaching of Mathematical Modelling, *op. cit.*, pp. 131–142.

Hall, G. G. (1984). The Assessment of Modelling Projects, *ibid*. pp. 143–148.

James, D. J. G. & Wilson, M. A. (1983). *Bull. IMA,* **19**, 9/10, 180–182.

McLone, R. R. (1979). *Bull. IMA*, **15**, 10, 244–246.

Oke, K. H. (1980). *Int. J. Math. Educ. Sci. Technol.*, **11**, 3, 361–369.

Oke, K. H. (1981). *Minimisation of Sound Distortion in a Record Player*. In R. Bradley, R. D. Gibson, & M. Cross (eds), *Case Studies in Mathematical Modelling*, Pentech Press, pp. 31–55.

Oke, K. H. (1984a). Mathematical Modelling—A Major Component in an M.Sc Course in Mathematical Education. Proceedings of the International Conference on the Teaching of Mathematical Modelling, *op. cit.*, pp. 86–95.

Oke, K. H. (1984b). Mathematical Modelling Processes: Implications for Teaching and Learning, PhD Thesis, Loughborough University of Technology.

Open University (1977). *Modelling by Mathematics TM 281*.

Open University (1982). *Mathematical Models and Methods MST 204*.

Treilibs, V. (1979). Formulation Processes in Mathematical Modelling, MPhil Thesis, University of Nottingham.

5

Formulation–Solution Processes in Mathematical Modelling

K. H. Oke
Polytechnic of the South Bank, London, UK

and

A. C. Bajpai
Loughborough University of Technology, UK

SUMMARY

This chapter reports on research that has been carried out on the nature of the complex linkages between formulation and solution in mathematical modelling. The processes involved in modelling are usually portrayed in a flow-diagram, or similar representation, and imply a *linear* or *linear with looping sequencing of stages*. Only recently, drawing on the still developing work of systems analysis (in information processing), have *nonlinear* approaches been suggested, and then only in the very broadest of terms. In order to try to understand more fully the highly complex processes involved, two theoretical constructs have been devised: a *concept matrix*, and a *relationship level graph*. The chapter defines and illustrates the ideas involved, first in general terms and secondly in terms of the initial attempts at modelling a problem (on central-heating) from scratch. The analysis involved, together with results of students' attempts at modelling a selection of problems, covers the following main points: *concept matrix portrayal of characteristics*—distribution of questions, assumptions, variables and constants, relationships between variables and constants; *relationship level graph showing how formulation–solution takes place*—basic (fundamental) relationship generation, forms of relationships, relationship 'level' as goal seeking, generation of variables and constants, and sub-problem emergence.

Finally, the implications of this work for learning mathematical modelling skills are examined. The discussion concentrates on the development of learning heuristics which are intended to offer some general guidance to students who are inexperienced in modelling. In the previous chapter the authors discuss a 'credit list' for use in assessment which is based on this work.

1. INTRODUCTION

It is now a common consensus that the *formulation* activities of mathematical modelling are the most difficult, Berry & O'Shea (1982), Burghes and Huntley (1982), Burkhardt (1981), James & Wilson (1983), Oke & Bajpai (1982), Penrose (1978), Treilibs (1979). Making sense of a practical problem and then making appropriate assumptions which lead to a set of tractable mathematical equations is a highly intuitive process. Since the start of a 'solution' to a problem will depend on the initial formulation, then it is the *formulation–solution interface* which is critical and poses the greatest challenge. The nature of the complex linkages between formulation and solution will be illustrated later in this chapter.

A survey of the leading methodologies of modelling, see for example Bajpai *et al.* (1982), Burkhardt (1981), Penrose (1978), Rivett (1980), Treilibs (1979), shows that each represents the processes of modelling either as a *linear sequence* of activities, as shown in Fig. 1, or the processes are represented by a *linear sequence with looping*, as shown in Fig. 2.

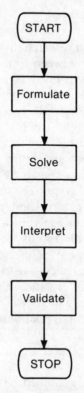

Fig. 1. Processes of modelling: linear sequence.

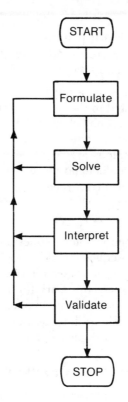

Fig. 2. Processes of modelling: linear sequence with looping.

Both Figs 1 and 2 are simplifications of the actual methodologies and are intended to show overall features only. Burkhardt (1981) and Treilibs (1979) have further analysed the formulation/solution activities, still based on a linear sequence (with looping). Treilibs has provided a further break-down of formulation and refers to his list as a set of skills:

  GV: generating variables
  SV: selecting variables
    Q: identifying the specific questions
  GR: generating relationships
  SR: selecting relationships

Clements (1982) suggests an alternative to linear sequencing in developing a framework of modelling processes. His development relates to the whole range of modelling activities, and draws its inspiration from the system movement of Checkland (1975). Clements quotes Checkland in referring to the distinctive, and nonlinear, features of the system approach:

> although the methodology is most easily described as a sequence of phases, it is not necessary to move from phase 1 to phase 7: what is

important is the content of the individual phases and the relationship between them. With that pattern established, the good systems thinker will use them in *any order, will iterate frequently*, and may well *work simultaneously on more than one phase*. (Our emphasis.)

So, Checkland is discussing a much more complex linkage of phases (or stages) than is suggested in the usual descriptions of modelling processes. A broadly similar philosophy of approach is adopted in this chapter, where the essentially nonlinear and holistic nature of formulation–solution processes is demonstrated.

In order to try to understand more fully the highly complex processes involved, two theoretical constructs have been devised, namely:

a concept matrix (CM)
a relationship level graph (RLG)

The next section of the chapter defines and illustrates the nature of the ideas involved. A subsequent section reports on the results of using CM and RLG in the analysis of a selection of students' attempts at modelling. Finally, the implications of this work for learning mathematical modelling skills are examined. The discussion concentrates on the development of learning heuristics which are intended to offer some general guidance to students who are inexperienced in modelling.

2. THE CONCEPT MATRIX AND RELATIONSHIP LEVEL GRAPH

The concept matrix (CM) arises from analysis of modelling activities and is designed to show which features, or concepts, are used in different modelling stages. The matrix is also intended to provide information on the type of each concept. Since the features which arise in the development of a mathematical model are extremely varied, both in clarity and in complexity, it was considered inappropriate to attempt to develop a simple hierarchy of concepts, as discussed for example by Skemp (1979). Initial attempts at classifying concepts by their relevance to the model were abandoned, since relevance only becomes clear in an *a posteriori* sense, that is after the model has been constructed and interpreted.

The relationship level graph (RLG) is designed to show that mathematical solution and formulation are interwoven; additional ideas on the nature of the problem are generated as a mathematical solution is developed. Initial, and more or less obvious simple relationships are denoted by the level 0 (zero). These relationships, although usually mathematical in nature, require no mathematical solution techniques to derive or form; they are mathematical representations of one variable and its dependence on another or others, written down from an *initial understanding* of the problem. This initial understanding, which might well arise from inspired guessing, is often related to knowledge of a non-mathematical type, for example of physics, biology, or medicine, depending on the problem. Usually, one, two, or at most three level 0 relationships need to be formed in order to be able to use mathematical techniques to form new relationships.

2.1 Concept Matrix (CM)

As pointed out earlier, the purpose of a CM is to show the nature of the features or ideas involved, ranging from the initial thoughts on a problem to the final stages of solution and interpretation. All key considerations as a model is being developed are entered in the matrix, their position being determined by how specific they appear to be (*specificity level*) and by their complexity (*complexity level*). These features or considerations are defined to be those statement, sketches and diagrams that consist of:

Questions
Assumptions
Variables and constants
Relationships between variables and constants

The matrix finally devised is two-dimensional and is represented in Fig. 3.

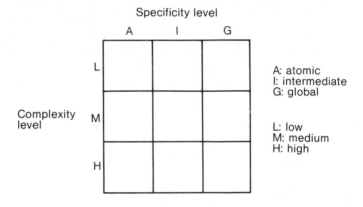

Fig. 3. Concept matrix (CM).

Later sections show that initial formulation takes place by identifying features that tend to fit at or near the bottom right-hand corner of the matrix. Early, and subsequent, solution activities involve features tending towards the upper left-hand corner of the matrix. Global features tend to be those that are only broadly related to the problem in hand, whereas Atomic features are those in the most simple form, for example variables or constants, which are immediately amenable to mathematical treatment. A High complexity level denotes a feature, that may be highly specific to the problem, but which may not be easily symbolised or quantified. Low complexity level indicates fairly easy quantification.

2.2 Relationship level graph (RLG)

The purpose of a RLG, as mentioned earlier, is to show that mathematical solution and formulation are interwoven. Initial understanding of the problem leads to simple relationships based on knowledge, guessing or both of the background to the problem. These first relationships are defined to be at level 0 (zero). The relationships deduced mathematically from level 0 are defined to be at level 1. After further mathematical solution work, and frequently the need for forming another level 0 relationship, level 2 types may be derived, and so on. Many modelling problems carried out by undergraduates, and others, reach a very significant stage by the time levels 6–8 are reached. A typical RLG showing relationship levels and their generation is shown in Fig. 4.

The number in each circle indicates the order in which each relationship is formed. A glance at Fig. 4 shows that there is no particular order in which relationships are formed, but that level numbers 0, 1, 2, . . . indicate overall progress from starting a model (relationships 1, 2 is level 0) to finish (relationship 16 is level 5). Note also that not all relationships generated are used in obtaining a final solution; for instance, relationships 10 and 11, level 2, are not used in obtaining 16. One of the most important features illustrated in a RLG is that the mathematical solution stage is intimately interwoven with the formation stage; mathematical techniques are themselves used in the generation of relationships. Most of the reported literature emphasises the need to formulate (generate features and relationships) *before* attempting a mathematical solution, although Burkhardt (1981), Treilibs (1979), and others have made the point that movement between formulation and solution is highly oscillatory. The

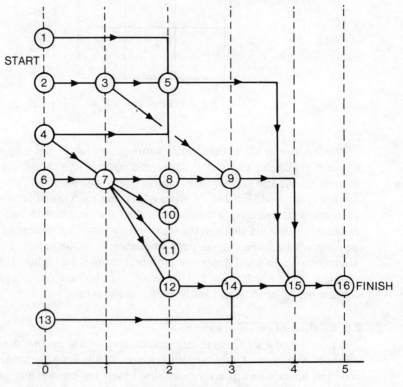

Fig. 4. Relationship level graph (RLG).

RLG shows, however, that formulation–solution processes are more complicated than a linear sequence of steps followed by oscillations. The numerical ordering in the circles shows an almost random order of events in some circumstances, whilst the generation of some relationships, for example 15 in Fig. 4, take place simultaneously working at a variety of levels (5, level 2; 9 and 14, level 3). The latter phenomenon is a clear illustration of Checkland's reference to the distinctive and nonlinear nature of the systems approach.

In the next section, in order to exemplify the characteristics discussed above, an analysis of one of the author's (Oke's) attempts at tackling a central-heating problem will be provided.

3. ILLUSTRATION OF USES OF CM AND RLG

The problem attempted is as follows.

> In using a central-heating system in a house, which is the best strategy for minimising heating costs.
>
> (a) Let house cool down naturally, central heating switched off when warmth not required.
> (b) Set thermostat to a certain value so that the house cools less when warmth is not required.
>
> Strategy required for any 24-hour period in winter.

Outline of modelling approach

To simplify the problem it is assumed that there is only one warmth period, during the day, and that there is only one cooling period, during the night, throughout any 24-hour interval. Furthermore, a very rapid response is assumed, so that the instant the central-heating boiler lights up, the heat it generates is immediately imparted to the house via the radiators.

The basic approach adopted was to consider Newton's law of cooling for the night and a net heat input equation (heat from system less heat loss = net heat gain) for heating up and maintaining a steady temperature. An expression for the difference in running costs between letting the house cool down naturally and keeping the night temperature above the minimum (which would have been achieved with natural cooling) is the end point of the exercise.

In order to illustrate as closely and as accurately as possible what happened in Oke's first crude attempts (of 6 hours' duration: first stab 2 hours, second stab 4 hours), the features considered below are in the order in which they occurred. All blind-alleys (e.g. solution paths dropped at intermediate stages) and groping around (what relationships to use, or derive, to do what and next) are included.

In order to save space only the relationships generated will be listed, although a small sample of features from the concept matrix will also be shown.

Relationship list

Relationship		Number (in order of occurrence)
Level 0	$H_1 = \dfrac{A\kappa}{d}(\theta_i - \theta_0) = K(\theta_i - \theta_0)$ (heat loss)	1
Level 0	$H_G - H_1 = 0$ (steady temperature: heat gain = heat lost)	2
$2 \to 3$	$\mathscr{C} = \pounds NKC(\theta_i - \theta_0)$ (Cost, steady temperature)	3
Level 0	$T\dfrac{d\theta_i}{dt} = H_G - K(\theta_i - \theta_0)$ (heating up)	4
Level 0	$T\dfrac{d\theta_i}{dt} = -K(\theta_i - \theta_0)$ (cooling down)	5
1 & 4 $\to$ 6	$t = \dfrac{1}{A}\ln\left[\dfrac{B - A\theta}{B - A\theta_i}\right]$ (heating up time)	6
$6 \to 7$	$\mathscr{C} = H_G t C$ (Cost of heating up in time t)	7
1 & 3 $\to$ 8	Heat loss = heat generated (night temperature $\theta_c > \theta_{min}$) $= K(\theta_c - \theta_0)(t_{1b} - t_4)$	8
Level 0	Heat generated = $H_G(t_{1b} - t_{1a})$ (temperature allowed to fall to θ_{min})	9
8 & 9 $\to$ 10	Difference in cost = $\pounds C$(RHS of 8 − RHS of 9)	10
$6 \to 11$	$t_{1b} - t_{1a} = \dfrac{1}{A}\ln\left[\dfrac{B - A\theta_{min}}{B - A\theta_c}\right]$	11
Level 0	Heat gained by house = $T(\theta_c - \theta_{min})$	12
$5 \to 13$	$t = \dfrac{1}{A}\ln\left[\dfrac{\theta_r - \theta_0}{\theta_c - \theta_0}\right] \neq \mathfrak{D}$	13
$13 \to 14$	$\theta_i = (\theta_r - \theta_0)e^{-At} + \theta_0$	14
$13 \to 15$	$t_{1a} = \dfrac{1}{A}\ln\left[\dfrac{\theta_r - \theta_0}{\theta_{min} - \theta_0}\right]$	15
$13 \to 16$	$\theta_c = (\theta_r - \theta_0)e^{-At_4} + \theta_0$	16
$13 \to 17$	$\theta_{min} = (\theta_r - \theta_0)e^{-At_{1a}} + \theta_0$	17
14 & 17 $\to$ 18	$\theta_i - \theta_{min} = (\theta_r - \theta_0)(e^{-At} - e^{-At_{1a}})$	18

Formulation–Solution Processes in Mathematical Modelling 69

12 & 18 → 19	Heat gained by house $= T(\theta_c - \theta_{min})$ $= H_G(t_{1b} - t_{1a})$ $\quad + (\theta_r - \theta_0)[T(e^{-At_{1b}} - e^{-At_{1a}})$ $\quad + Ke^{-At_{1a}}(t_{1b} - t_{1a})]$	19
11 & 17 & 19 *→ 20*	$T(\theta_c - (\theta_r - \theta_0)e^{-At_{1a}} - \theta_0) = $ RHS of 19	20
8 & 13 & 20 *→ 21*	Solve for t_{1a}, t_{1b}	21
20 → 22	Difference in heating amounts for given θ_c	22
10 & 21 & 22 *→ 23*	Difference in costs in terms of θ_c	23

The following should be noted.

(1) The relationships list is provided to indicate for each relationship:

    (i) its form,
    (ii) other relationships from which it is derived, for example 8 & 9 → 10 implies that relationships 8 and 9 are used to derive relationship 10.

(2) The concept matrix (CM) shows each feature in order of occurrence (A–Z, AA, BB, ...). Where letter(s) and a number appear together then the feature is a relationship; for example (HH20) means that feature HH is relationship 20. (See Fig. 5.)

Other examples of features are:

A Thermal capacity of system (house) (T)
B Heat generated by boiler and radiators

.

.

.

E Heat loss is involved

.

.

.

G Cost of maintaining a particular temperature is needed

(3) The relationship level graph (RLG) shows all relationships from the list. Note that relationships 1, 2, 9, 4, 5, 12 are each of level 0; these relationships require no mathematical derivation and depend solely on interpretation of the problem statement and associated basic physics. Relationships at level 1 and above are derived mathematically, for example relationship 3 is derived from relationship 2 (level 0) and

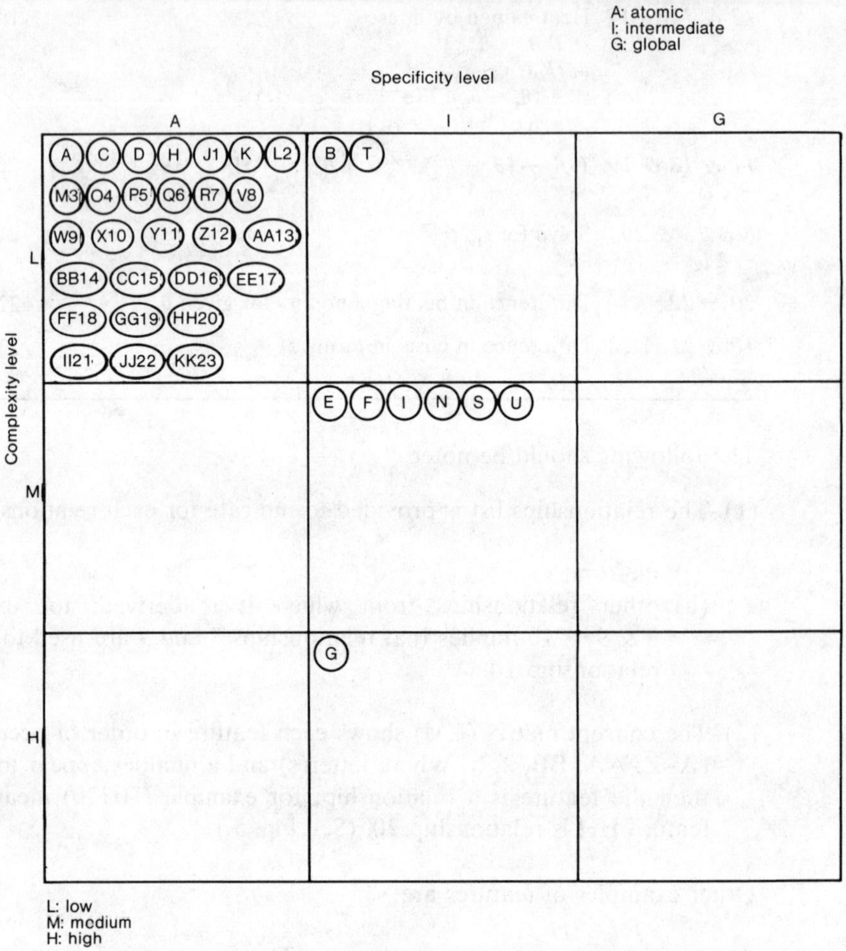

Mathematical Modelling—Methodology, Models and Micros

A: atomic
I: intermediate
G: global

Fig. 5. Concept matrix: central-heating problem: Oke's initial modelling attempts.

L: low
M: medium
H: high

hence relationship 3 is at level 1. *Intermediate mathematical detail in deriving a relationship is not shown* (see Fig. 6).

(4) An element of subjectivity is inevitably involved in the construction of the CM and RLG, particularly in the former. However, a number of colleagues have constructed both for the problem (and other problems) and close agreement has been observed in each case.

The points, numbered (1)–(4) above, are general and refer to the essential characteristics of an analysis of any modelling attempt using a CM and RLG, no matter what the original problem. The following are interpretations of Figs 5 and 6 and thus relate specifically to Oke's initial attempts at modelling the central-heating problem. However, several of the interpretations have a wider significance for modelling processes in general and these are examined as they arise.

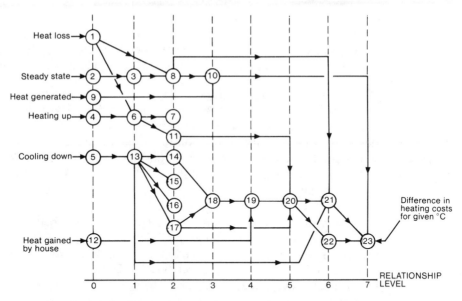

Fig. 6. Relationship level graph for central-heating problem: Oke's initial modelling attempts.

I *Distribution of features in CM* Most of the features, even the ones identified initially, tend to be highly specific to the problem and also tend to be the most easily quantified (at least in principle). Hence the cluster of features in the top L-H corner and the sparsity of features in the other squares that is the most noteworthy characteristic, since relationships are defined in general to fit in the top L-H corner.

II *Generation of variables and constants* Variables and constants are largely generated as relationships are formed. An analysis of the features involved shows that out of 15 symbols (variables and constants) generated, only three (θ_i, θ_0, T) were thought of before relationships were formed. An additional five symbols were introduced in level 0 relationships, an additional three symbols at level 1, and the final four at level 2; the last symbol to be introduced is θ_{min} which occurs in relationship 11. Symbols such as A and B which were introduced solely for mathematical convenience are not included.

As to be expected, towards the end of the goal seeking (high relationship levels), symbols of prime importance to the problem are no longer generated.

III *Level 0 Relationships* The hardest part in getting started with any modelling problem is the formation of the first level 0 relationships. In this case, relationships 1 and 2 provide the starting point. Experience in modelling, as well as in the problem class (elementary heat exchange), appear to be important factors which lead to improvement. As the solution progresses, however, additional insights are gained and these prompt the need for further information. Hence the generation of relationships 4, 5, 9 and 12, each at level 0. *Mathematics* (solution) has helped in the *intuitive* (level 0) understanding of the posed problem.

IV *Formation of relationships at levels 1, 2, ...* Relationships are often generated by working simultaneously at a variety of levels, for example relationship 19 (level 4) is formed from 18 (level 3) and 12 (level 0). Note that not all relationships generated at a given level are subsequently used, for example relationships 7, 15, 16, (all at level 2) make no contribution to the solution (relationship 23) and therefore are redundant.

V *Sub-problem identification* The RLG is partitioned into two distinct regions as far as relationship 19 (level 4); the upper region starts with relationships 1, 2, 9, 4 and the lower region starts with relationships 5 and 12, all at level 0. Not until relationship 20 (level 5) is reached, is a link formed between the two regions. The upper concentrates mainly on heating up, and the lower mainly on cooling down of the house. Each region therefore represents the development of a sub-problem, where the two sub-problems are combined at relationship 20. The author (Oke) was totally unaware whilst modelling that these two sub-problems were in fact being tackled; it felt like working on one problem only.

With some problems, however, it is not only possible to identify sub-problems at the outset, but it is quite clear that the problem can be broken down into very distinct parts. For example, when considering the shape and size of a pick-up arm in order to minimise sound distortion in a record-player (Oke, 1981a), it is clear *before any mathematics is attempted* that the following are key sub-problems:

(a) minimisation of sound distortion from a purely geometrical approach (treat recording groove as a system of concentric circles, concentrate on angle between a tangent and the arm at any point);
(b) minimisation of sound distortion from a signal analysis approach (consider a sinusoidal transverse wave and the advance and retard effect produced by a non-tangential arm).

4. RESULTS OF USING CM AND RLG IN THE ANALYSIS OF STUDENTS' MODELLING ATTEMPTS

In order to test the appropriateness of CM and RLG, a selection of students' attempts at modelling was analysed. The problems concerned and the type of student involved are summarised in Table 1.

The references after each problem contain details on possible modelling approaches. Briefly, the baby's milk bottle problem involves placing a baby's bottle full of cold milk into a saucepan of cold water and then placing on maximum heat on a cooker in order to heat up the milk to blood

Formulation–Solution Processes in Mathematical Modelling

Table 1. Problems and student types

Problem	Type of student
Modelling the heating of a baby's bottle (Oke, 1979)	MSc Math. Ed.
Minimisation of sound distortion in a record player (Oke, 1981)	BSc Appd Physics (2nd year)
Speed-wobble in motorcycles (Oke, 1981)	MSc Math. Ed.
Evacuation of a school (Wilson, 1983)	First year sixth form, secondary school

temperature in a minimum time; how much water should there be in the saucepan? The record-player problem has already been referred to. Speed wobble refers to oscillations in a castor: what causes this phenomenon? Evacuation of a school involves the problem of what order should classrooms be evacuated, and total school evacuation time, in the event of a fire. The problems were chosen for the variety of potentially different modelling approaches that were possible (although they are all analytical and deterministic rather than stochastic). Regarding the types of students who tackled these problems: MSc Mathematical Education is a two-year part-time course for graduate teachers of mathematics in secondary schools and colleges of further education; most have a background knowledge of physics to approximately GCE A-level (although rusty). Students on the BSc Applied Physics course have completed one year programmes of mathematics and physics. The first year sixth formers were all taking pure mathematics at GCE A-level. *All the students had little or no experience of mathematical modelling.*

The students were split into groups of approximately four in each case and had only the relatively short duration of 3–10 hours (split over several sessions) in which to work. Each group was required to keep a careful log of all initial thinking, working and scrap-work. Each log of each group (a group log rather than individual written work was accepted for the purpose of these experiments) was analysed using a CM and RLG. Lecturer hints throughout each exercise were kept to a bare minimum, and were given only to prevent frustration or fixation (of ideas); even then, information was provided only in a broad form, for example, 'try some algebra and drop scale-drawings'.

Overall, since the experiments were of short to medium duration and also because the students referred to had little or no modelling experience, the RLGs were considerably less developed than the one illustrated in Fig. 6. For the same reasons, fewer features appear on each CM. In order to give an indication of the student's work, a representative RLG for each problem is outlined in Fig. 7 (to save space each RLG is drawn to approximately one-third of size of Fig. 6, that is, to one-ninth of area).

The speed-wobble problem was found to be the hardest in the time available (approximately 3 hours), and consequently the RLG shown in (c), Fig. 7, shows little structure. The overall features of a RLG, as discussed in the last two sections, are also illustrated in Fig. 7. In particular, two

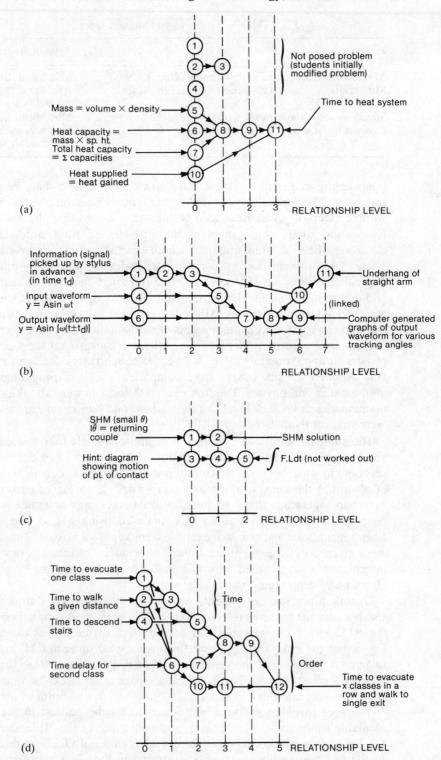

sub-problems are identified in (d) on the evacuation of a school problem; neither the students nor their teacher (Wilson) had thought of the two problems at the outset, namely, that time and order in which classes evacuate into a corridor were two distinct aspects.

The CM and RLG, particularly the latter, help in showing how formulation and solution take place in modelling processes, and they are capable of being used in the analysis of a variety of different students' attempts at modelling (from school to HE). By comparing the RLGs of each group working on the same problem, student performance can be most readily compared (this has been done by the authors although, as explained earlier, only one group per problem is illustrated in Fig. 7).

There are, however, two outstanding matters that require further investigation. First, the creative leap that is needed in the formation of the first level 0 relationship (*before* any mathematics is carried out) is little understood. Clearly this is a very difficult matter and all that can be said for the present is that students improve, as with modelling expertise generally, with more practice. It also seems very important for students to gain practice by modelling a particular class of problems where common features arise. Secondly, the strength or importance of relationships, apart from the basic or level 0 type, also needs investigating. Deeper insights into the direction of main thrust of formulation and solution would no doubt accrue if such strengths could be defined.

5. IMPLICATIONS FOR LEARNING MATHEMATICAL MODELLING

This section concentrates on learning mathematical modelling. The implications of this work for assessment are examined in the previous chapter. Discussions on teaching styles and the overall place of mathematical modelling in a curriculum may be found in a number of publications, for example Burghes & Huntley (1982), Burkhardt (1981), Oke & Bajpai (1982).

Learning heuristics

In an attempt to help inexperienced students in modelling, and in the spirit of offering general guidance in the hope of providing some confidence in what is an unfamiliar activity, a list of heuristics (rules-of-thumb) was devised. The construction of the list was based partly on published literature in problem solving and mathematical modelling, and partly on the work of the previous sections on formulation-solution processes. The published literature may be referred to in Polya (1957), Kilpatrick (1969)

Fig. 7. Relationship Level Graphs (RLGs) of a selection of students' attempts at modelling (a) Heating of a baby's milk bottle: MSc Math. Ed., group 1. (b) Minimisation of sound distortion in a record player: BSc 2 Appd. Physics, group 4. (c) Speed-wobble in motorcycles: MSc Math. Ed., group 4. (d) Evacuation of a school: first year sixth form (Wilson, 1983).

and Gagné (1966), for example, in problem-solving. In the case of mathematical modelling processes, although the term heuristics is rarely used, reference may be made for example to Morris (1967) (early but very helpful work), Bajpai *et al*. (1982) and Burkhardt (1981).

The number of heuristics in the list has been kept deliberately low. The reasons for this are:

(a) too many considerations serve only to confuse when considering any one problem,
(b) a large list would tend to make each heuristic highly specialised and so dependent on a specific problem being considered.

The list, which aims to cover most of the initial stages of modelling activity, is as follows.

(i) Establish a clear statement of objectives.
(ii) Do not write a vast list of features.
(iii) Simplify (build up gradually, make guesses).
(iv) Get started with mathematics as soon as possible.
(v) Carry out some mathematics on initial relationships.
(vi) Got a solution yet? (If not simplify the maths.)
(vii) Know when to stop. (Do not seek perfection.)
(viii) Interpret your solution.
(ix) Validate your solution.
(x) If stuck observe practical situation or carry out a 'thought experiment'.
(xi) Have frequent rests.

The work on formulation–solution processes described in the previous sections has been shown to support this choice of heuristics, in particular the following.

(i) *Establish a clear statement of objectives* See I (distribution of features in CM) and V (sub-problem identification) in Section 3. Encourage students to keep a log of all rough work done and to include initial 'vague' thinking; from this initial work, it is easier to get some reasonable objectives on how far to go, that is, what type of solution or solutions are being sought. Do not insist on initial partitioning of problem that is, identification of sub-problems; the partitioning might well evolve naturally at a later stage (of the formulation–solution process).
(ii) *Do not write a vast list of features* With experience, students appreciate the virtue of this heuristic. See I, II (generation of variables and constants) and III (level 0 Relationships) in Section 3, which show that additional features are identified as the solution is developed.
(iii) *Simplify* Build up very gradually. Make guesses, make assumptions, add restrictions. Lump components (attributes) together and treat as single component (see Figs 6 and 7).
(iv) *Get started with maths as soon as possible* Identify a few variables, parameters, constants. Write down one or two obvious mathematical

relationships. Keep mathematics as simple as possible. See I, II, III, Section 3.

(v) *Carry out some mathematics on initial relationships* This itself generates more variables, constants, and relationships. See I, II, III, Section 3.

The order in which the heuristics have been listed is not necessarily the order in which they may be recommended for use. The results of teaching and learning experiments, together with the formulation–solution analysis of the previous sections, show that modellers (experienced as well as inexperienced) move forwards, recap, then move forwards again often carrying out several modelling activities simultaneously. However, the main intention of the heuristics is to provide some sort of guidance for the inexperienced when a new problem is starting to be tackled. In which case, the first few heuristics might with advantage be carried out in the order listed, that is, starting with (i): 'Establish a clear statement of objectives' and working to (v): 'Carry out some mathematics on initial relationships'.

In an attempt to gauge student opinion of the usefulness of the heuristics as initial guidance in modelling, a questionnaire containing the full list, (i)–(xi) was issued. Students were asked to rank the usefulness of each heuristic according to a numerical scale: essential (value 1) down to useless (value 5). Several undergraduate and MSc Math. Ed. classes, including the groups reported on in Section 4, were issued with the questionnaire. For each class, the average rank for each heuristic was calculated. Overall, students found the heuristics useful as measured by the grand average rank for all eleven heuristics (value: 2.6). There is inevitably considerable variation amongst individuals and groups of individuals in the ranking value (1–5) given to each heuristic, but the most popular (useful) heuristics are chosen by most as (i), (iii) and (iv). As experience in modelling is gained, students realise the value of each heuristic and give each a low (very good) ranking.

6. CONCLUSIONS

Two theoretical constructs, namely a Concept Matrix (CM) and a Relationship Level Graph (RLG), have been devised and used in the analysis of formulation and solution processes in a range of problems. The discussion in the preceding sections has illustrated the complex nature of the processes involved and has shown that formulation and solution are intimately interwoven. The relationship level graph, in particular, has shown that much of the modelling process is nonlinear in nature and that several activities are often carried out by working at a variety of stages simultaneously. The emphasis throughout has been on students who are inexperienced in modelling and who have in general had only a short time in which to tackle the problems involved. This latter constraint is considered to be realistic in view of the usual pressures which exist in an educational environment (except where extended project work, for example end-of-course assessment, is involved.

The following are the main points that have emerged.

(1) *Distribution of features* There is no discernible order in which features are recognised although there is a general movement from the bottom R-H corner of the concept matrix in early stages to the top L-H corner at the onset of a solution.
(2) *Basic relationships are often generated as solution proceeds* The mathematical solution itself helps with further understanding and hence formulation of the problem by prompting the need for level 0 relationships.
(3) *Relationships can occur in various forms* General: (L, I) position in CM. Applicable not only to problem in hand (e.g., mass = volume x density, see (a), Fig. 7) Specific: (L, A) position in CM. Directly related to problem.
(4) *Relationship level as goal seeking* As with features generally, relationships often occur in no discernible order. However, a measure of the general progress made in finding a solution is provided by relationship level.
(5) *Most variables and constants are generated with relationships* As mathematical deductions are made in the generation of relationships, so variables and constants are more naturally introduced.
(6) *Sub-problem identification* It is difficult to find a general rule regarding the recognition of sub-problems. Sometimes sub-problems are identified at the outset, on other occasions they are only recognised by partitions in a relationship level graph.

In future work it is intended that students are also encouraged to construct their own concept matrix and relationship level graph *as their problem proceeds*, in order to see if they gain further understanding of their attempts and thereby improve their solution as it develops. It is predicted that blind-alleys (manifested in redundant relationships) and sub-problem partitioning may become apparent early in the modelling and so clearer thinking will ensue which in turn could lead to less laborious solution paths.

The implications of the formulation–solution analysis in the development of learning heuristics are discussed in Section 5. Eleven heuristics have been developed and overall student opinion has supported the usefulness of each.

REFERENCES

Bajpai, A. C., Mustoe, L. R. & Walker, D. (1982). *Engineering Mathematics*, 2nd edn. John Wiley & Sons.
Berry, J. & O'Shea, T. (1982). *Int. J. Math. Educ. Sci. Technol.*, **13**, 6, 715–724.
Burghes, D. & Huntley, I. (1982). *Int. J. Math. Educ. Sci. Technol.*, **13**, 6, 735–754.
Burkhardt, H. (1981). *The Real World and Mathematics*, Blackie.
Checkland, P. B. (1975). The development of systems thinking by systems practice—a methodology from action research programme. In R. Trappl, F. de P. Hawika (eds). *Progress in Cybernetics and System Research*, Vol. II, Hemisphere, Washington DC.
Clements, R. R. (1982) *Teach. Maths. & Its Applcns*, **1** 3, 125–131.

Gagné, R. M. (1966) *Human problem solving: internal and external events*, in Kleinmuntz B. (ed.), *Problem Solving: Research Method & Theory*. John Wiley & Sons

James, D. J. G. & Wilson, M. A. (1983). *Bull. IMA*, **19**, 9/10, 180–182.

Kilpatrick, J. (1969). *Rev. Educ. Res.*, **39**, 4, 523–534.

Morris, W. T. (1967). *Management Science*, **13**, 12, B707–B717.

Oke, K. H. (1979). *Int. J. Math. Educ. Sci. Technol.*, **10**, 1, 125–136.

Oke, K. H. (1981a). Minimisation of sound distortion in a record player. In R. Bradley, R. D. Gibson & M. Cross, (eds). *Case Studies in Mathematical Modelling*, Pentech Press.

Oke, K. H. (1981b). Speed-wobble in motorcycles. In D. J. G. James & J. J. McDonald, (eds). *Case Studies in Mathematical Modelling*, Stanley Thornes.

Oke, K. H. & Bajpai, A. C., (1982). *Int. J. Math. Educ. Sci. Technol.*, **13**, 6, 797–814.

Oke, K. H. (1984). Mathematical Modelling Processes: Implications for Teaching and Learning, PhD Thesis, Loughborough University of Technology.

Penrose, O. (1978) *JMMT*, **1**, 2, 31–42.

Polya, G. (1957). *How to Solve It*, 2nd edn (still in print). Doubleday Anchor Books.

Rivett, P. (1980) *Model Building for Decision Analysis*. John Wiley & Sons.

Skemp, R. J. (1979). *Intelligence, Learning and Action*, John Wiley & Sons.

Treilibs, V. (1979). Formulation Processes in Mathematical Modelling. M. Phil. Thesis, University of Notthingham.

Wilson, D. B. (1983). The Promotion of Mathematical Modelling and Mathematical Understanding in Schools, MSc. dissertation, Polytechnic of the South Bank (CNAA).

6
A New Approach to Model Formulation

D. E. Prior
Sunderland Polytechnic, UK

The expression of any experience is first and foremost through word association. The richness of a person's language contributes directly to that person's ability to express the factual and the mood content of a 'real world happening', both capturing the moment for himself and also allowing him to share it with others. Those with whom the observer wishes to share this experience must be capable of interpreting the verbal utterances in such a way that the essential experience of the observer is preserved. The above simple observation must have profound consequences in the world of modelling since all models stem from the abstraction from the world at large of an experience by a modeller and no matter how the model ultimately is represented it begins life in mental imagery expressible only, in the first instance, in and through words. This 'embryonic model' stage in the construction of a model has had very short shrift from modellers generally and the mathematical modeller in particular and yet it is a most vital link in the modelling process, bridging, as it does, the two worlds of experience and model construction.

The following chapter is an attempt to show that the modeller may take an alternative path from the 'real' to the 'model' world to that which usually is mapped out for him on account of the chosen method of solution. Attention is drawn to the fact that this chapter does not concern itself with the full methodology of model formulation developed at Sunderland, only that part of it which has been found in practice to be the most difficult to execute. The content of the chapter is based upon a session carried out with a group of experienced modellers who agreed to act as modelling guinea pigs.

Before any account is detailed of experiments entailing new approaches to the formulation of a model the concept of 'model formulation' in the context of this chapter first should be made more clear, and some discussion ensue concerning a popular approach to model conceptualisation and formulation.

Any experience, or system, in which the modeller has an interest is seen by him to contain certain discrete elements which are perceived to bear relationships each to the others and which altogether, through some process applied by the modeller, become the 'model' for the experience. So far as this chapter is concerned the process of model formulation is that process which yields the all important set of elements and does not necessarily yield the relationships between them.

Once the relationships between the elements are expressed then the problem, as such, is 'solved', or, rather, 'fixed'. Expressing particular relationships between elements through equations and solving such equations is not necessarily solving the problem. When the modeller speaks of 'solving' he means 'deeper, or fuller, understanding of'. Equation solving is purely a mechanical process which does nothing more than produce a certain numerical result which is intrinsic in the expression to start with.

Although what is said above is neither particularly deep philosophically nor complex intellectually the problem is that many modellers, and most mathematical modellers, appear to miss, or choose to ignore, the point and seem, on the evidence of their current performance, convinced that the model formulation stage is the immediate production of an equation or set of equations. Modellers take for granted that the elemental content of problems is not a difficult issue provided the 'correct' standpoint from which to view the problem has been decided upon and tend to concentrate their efforts entirely upon determining what type of relational operators need to be applied to the set of variables chosen to represent the elemental content so that the problem may be 'solved'. That this is so is found by consideration of a most widely used form of model formulation method called the *feature list* approach (Fig. 1). In the feature list approach all relevant features of the system, or experience, are written down by the modeller. By what process the modeller decides how this feature or that is or is not relevant is seldom discussed, but what is interesting to note is the process by which the modeller decides that a feature having been chosen is now no longer worthy of further consideration and should be struck off the feature list. This process of the removal of a feature from the list seems to depend upon whether or not its presence is embarrassing in that it hinders the tractability of an equation which the modeller somehow *seems to know beforehand that he is going to write*. In a nutshell *the model construction stage of the feature list approach is driven* from the solution regime. No wonder experienced modellers can get away with using the feature list approach whereas beginners get hopelessly lost. As the process unfolds the experienced modeller 'knows' or develops a 'feel for' the particular basic equation isomorph for the experience to be modelled and can therefore gradually manipulate the feature list, following a pseudo-form of 'eureka!' arguing, to produce the desired elemental content to which the requisite mathematical operators may be appended to produce the required, anticipated, solution.

What the above implies of course is that in reality most mathematical modelling is to do with shoe-horning the breadth of man's experience into

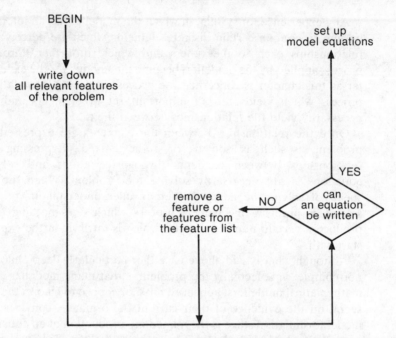

Fig. 1. Feature list approach to model formulation.

a very few well tried and tested equation types for which the solution techniques are well known. The mathematical modeller fools himself, and others, into believing that he is solving many and varied situations when in reality he is continually trimming down experience to conform to a few well understood patterns. Any creativity in the naive modeller is stifled by the construction of an edifice of standard techniques which eventually drives him to invert experience and formulation. Experiences stem from and are cast in the moulds of whatever techniques are selected to 'solve' the experiences.

The above approach can only lead to a race of modellers who *naturally and as a matter of course expect their world to be ordered* and who think *mechanically*, not *analytically*. Thus the experienced modeller says to the beginner who clearly is making no headway at all in the formulation of a model, 'never fear, with experience, the art of modelling will be acquired and you will find yourself modelling'. What the experienced modeller really means is 'never mind, once you have grasped the few standard analogies we use to model experiences around us you'll be all right'.

From the foregoing, however, no-one should construe that the benefits which can result from the proper application of the principle of isomorphism are not fully appreciated. The concept of isomorphism when used properly is a very powerful aid to modelling and to the further understanding of systems, but when used as many 'feature list' modellers use it then errors of model formulation can occur.

Before the group of modelling guinea pigs could settle down and the experiment get under way some discussion took place and an agreement

A New Approach to Model Formulation

was reached about the definition of a model. Definitions of a model and also of what constitutes exactly the modelling process abound in the literature, but most seem to miss the point in that there is an implicit assumption that the reader has sorted out the relationship between his environment and the verbal interpretation of the signals reaching him from that environment. In actual fact nothing could be further from the truth. However, it is not the intention to discuss here issues arising from the above concept, essential though it may be. Suffice it to say that discussion of these ideas was useful in arriving at a definition of a model which the group found most satisfactory for use during the experiment.

The group decided that the most important criteria by which a modeller might judge whether or not he has obtained a 'model' of the experience under scrutiny are that the model must be

(1) a representation of that experience,
(2) capable of being validated,
(3) capable of being used, by itself, for further study of that experience.

Having determined the operating standards for the experiment the group were then presented with the names of two common systems and asked to construct a model of each of them. The first system the group was asked to model was 'a pendulum' and the second system 'a fire'.

The pendulum presented no problems, most of the group returning the first of the following two mathematical models for the periodic time. One or two supplied the second, more complex, solution which holds for oscillations larger than those near which the former simple model breaks down.

$$T = 2\pi\sqrt{\frac{l}{g}} \quad \ldots \text{small oscillations} \quad (1)$$

$$T = 4K\sqrt{\frac{l}{g}} \quad \ldots \text{large oscillations} \quad (2)$$

where $\quad K = \int_0^1 (1 - t^2)^{\frac{1}{2}}(1 - k^2 t^2)^{-\frac{1}{2}} \mathrm{d}t$

The fire, however, caused a minor revolution. Numerous questions (and objections) were placed by the group in response to the request to model this concept. Typically responses ranged from the indignant

> you can't model a fire!

through

> insufficient information—what exactly is required—the heat produced by a fire at different distances from it, or the speed at which the fire is advancing—or what?

to

> If I am to model the chemical equations of combustion then at what depth of analysis am I to stop?

Overall it was obvious that the group had a definite, firm, clear conceptualisation of a pendulum, but felt stranded upon the slippery slopes of indecision and uncertainty so far as the abstraction of the concept of a fire from the real world was concerned. It is interesting to note that the questions about the fire hinged not at all upon the essence of 'fire' itself, or whatever concept that word may have provoked within the group's combined consciousness, but instead only upon the results consequent to a conflagration combined with familiar, well entrenched concepts of the modelling world namely, heat and a heat gradient, and speed of movement and an area extension. It was as though the group had subconsciously agreed to shun the unfamiliar and collectively rearrange the question posed to align upon more familiar principles and effects associated with, and consequential upon, the phenomenon of 'fire'.

So far as the pendulum was concerned no such confusion existed. The essence of a pendulum was embodied in the equations given above and that was that. Perhaps so far as the pendulum is concerned the members of the group all felt to be very much in the position of the experienced modeller alluded to at the beginning of this chapter, whereas with regard to the fire each member felt very much the beginner, having nothing at all to fall back upon as an analogue of a fire.

In order to help the group to recover their composure some points were put to them about the process of model abstraction and in particular the following two essential ideas floated. First, that no matter in what way the 'real world system' is ultimately represented or 'modelled'—be it schematically, iconically or mathematically—the process of abstraction from the world is by the application of many successive mental 'sieve and compress' processes each one filtering and moulding the conceptualisation of the experience *aligning the abstraction more and more upon the beliefs educated, instilled and subconsciously absorbed into the modeller through his cultural background and each one separating the experience more and more from the final model*. Secondly, and this is the more important and profound of the two points, the process of abstraction *takes over from that point in the process where the modeller first believes he has grasped a mental image of the system under study*. This last statement is tantamount to declaring that for the average modeller, once the perceived system and the idea (or mental image) of the system are linked then the process of model abstraction and construction is mechanical, limited only in scope by the technical expertise and experience of the modeller in making complex constructions analytically tractable. In other words, once the modeller thinks he has got it, so to speak, everything else that follows in his endeavour to model the experience is encompassed, limited and defined by his ability to apply a known solution technique.

The group then was encouraged to consider what might happen were the modeller to concentrate his attention upon that phase in the model abstraction process immediately preceding the mechanical phase discussed above and which is termed simply the 'verbal expression' phase to yield what was described at the very start of this chapter as the 'embryonic model'.

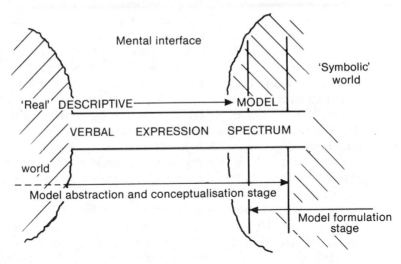

Fig. 2. Model abstraction and development through 'VESPA' (Verbal Expression Spectrum Approach)

A very simple, but quite effective, schematic representation of the verbal expression phase of use in encouraging beginning modellers to test their ability to abstract experiences from the world about them is illustrated in Fig. 2. As the statement describing the experience is studied and manipulated, attempts are made to 'push' the statement from the left-hand end of the spectrum where the sheer *descriptive* aspects are very much to the fore and the *model* aspects—as laid down by the earlier elementary criteria—are not in evidence, to the right-hand end where the *model* considerations are dominant and the *descriptive* side supportive rather than precipitating ambiguities.

The group now was redirected to produce two verbal representations of the experiences 'a fire' and 'a pendulum' which could be positioned upon the verbal expression spectrum. The following summarises the efforts of the group:

A FIRE
A pile of dry sticks is placed upon dry, crumpled paper and the paper ignited. When substantial flames are observed coal is heaped liberally over the pile of wood and the whole then left to burn.

A PENDULUM
A heavy weight, at rest, suspended by a light thread, is pulled aside a short distance from its point of rest and then released.

Note that no attempt was made at any stage to redirect the group's energies or attention through criticism of the initial representations. For example, it would have been very easy to comment that the statement for the fire was less a representation of 'a fire' and more a representation of 'making a fire'. Similar comment might have been directed to the initial representation of the pendulum. It has been discovered that most aspiring modellers hardly ever construct the 'essential' model, preferring instead to formulate a representation which covers a description of the content of the experience under scrutiny. The argument is always open, of course, as to which of the representations is the model—the representation of what the experience is through what it contains or through what it appears to do. Needless to say, to attempt even to begin to unravel the thread of this particular cocoon of model abstraction philosophy would absorb more time and energy than this chapter is allowed, yet the points are made to indicate to the reader that this part of the modelling methodology needs careful thought and discussion.

To return to the group experiment each statement was then tested against the three criteria for model acceptance. Each was certainly a representation of an experience. Each also could be verified by the simple expedient of following the words and returning from the statement back to the experience. Had the statement for the fire, say, used the word 'wet' instead of 'dry' then the statement soon would be shown to be invalid. However so far as the third criterion was concerned no member of the group sincerely could say that the statements aided further study and appreciation and understanding, through them, of the systems to which they referred. The statements were too far over to the *descriptive* end of the expression spectrum. According to the criteria the statements were much nearer descriptions of the named systems than models; but the statements were a start. The very fact that some agreement had been reached amongst the group members about a common statement which modelled, albeit weakly, their collective association of the concept of 'a fire' in the real world about them was an extremely important first step. No matter how weak the 'model' content of the statement it could now be operated upon to push it through the expression spectrum strengthening its 'model' content by seeking support from a more skilful and definitive arrangement of the 'descriptive' content. Such rearrangement is achievable through collaborative discussion. Greater insight is gained into the meanings of the terms used in the statements through sheer familiarisation with them over prolonged debate. In this manner does this part of the model conceptualisation method answer the above implied criticism of current modelling approaches that insufficient time is devoted to the familiarisation, insight gaining and understanding of the problem through discussion in the early phases of model construction?

In this manner then followed what was, for the group as a whole, a very frustrating hour. The members were asked continually to review the statement and as a result gradually 'push' it as far towards the *model* end of the spectrum as possible.

Finally the following statement for the verbal *model* of a 'fire' was obtained

A FIRE

A speed of combustion which always tends in a direction so as to meet the amount of combustible air, which always adjusts to oppose the direction of the speed of combustion.

The reader is invited first to re-read the earlier statement for the fire and compare it with the above and then to test the above statement against the criteria for model qualification and satisfy himself that indeed the statement has passed from one end of the expression spectrum to the other—from being a verbal description of 'a fire' to becoming a verbal model of one.

The next stage of the methodology depends upon to what extent the system under scrutiny is perceived as a feedback system. In this context feedback is broadly defined as the 'modification of input streams to the system by all, or parts of, output streams from the system'. If any feedback relationship along these lines is deemed to exist in the system, and is reflected in the model, then a well established technique is next employed to 'push' the model off the end of the verbal expression spectrum into a no-man's land which bridges model conceptualisation and model construction. This well known technique is the cinderella of the modelling world and is called causal loop diagramming. It is a perfect complement to the method of model conceptualisation through the verbal expression spectrum for feedback systems models, in that it literally shifts a feedback system model sufficiently beyond the rather arbitrary associations with which words endow even extreme right-hand end models of the verbal expression spectrum into an area where the aura of quantitative, as opposed to qualitative, judgements makes itself felt. Causal loop diagramming is the important conceptual link between the qualitative assertions of experiences and their quantitative equivalents which usually confound and confuse the less mathematically biased, but the nonetheless sincere, modeller.

Briefly, a causal loop diagram is a schematic representation of a verbal model which has been pushed to the extreme RHS of the verbal expression spectrum and in which certain elements of the model are connected by arrowed lines. Each arrowed line represents the influence that the tail element of the arrowed line exerts over the head element. Each arrow carries either a positive or a negative sign. A positive sign signifies that any change in the tail element of the influence link results in a change in the head element which is in the same direction as the change in the tail element. Thus a decrease in the tail element of a positively signed link results in a decrease in the quantity associated with the head element. Similarly an increase in the tail element results in an increase in the head element. A negative sign signifies that any change in the tail element of the influence link results in a change in the head element which is in the opposite direction to the change in the tail element. Hence a decrease in the tail element of a negatively signed influence (or causal) link results in an increase in the corresponding quantity at the head of the arrow. A decrease in the value of the tail element of a negatively signed influence link results in a consequential increase in the value of the head element of the link.

Using this technique the modelling group produced the causal loop model depicted in Fig. 3 of the previous verbal model of 'a fire'. As can be

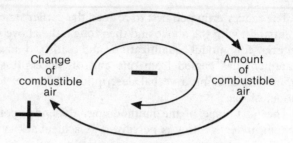

Fig. 3. Causal loop model of 'a fire'.

seen from the negative sign assigned to the single influence loop, the causal loop model allows the system depicted to be analysed further by its clear presentation of the nature of the feedback at work in the model, in this case negative or goal seeking. It is immediately evident that upon the basis of the model developed thus far a fire is a negative feedback system. In other words, counter-intuitively, the model is saying that fires do not burn more fiercely the more fiercely they burn, instead they continually seek to achieve a state of equilibrium in which the fierceness of the burn grows or diminishes to match the amount of burnable air reaching the area of the burn.

And what of the pendulum?

The reader is invited first to study the verbal statement on page 85 and then, before reading on, to draft for himself first a verbal model of 'a pendulum' and then to push this model off the end of the expression spectrum by constructing a corresponding causal loop model.

The results are, as the modelling group discovered, both startling and quite revealing in terms of the power of this method to provoke thought and discussion about what was hitherto believed to be a simple, straightforward and well understood system.

The modelling group finally produced the following statement for the verbal model of 'a pendulum':

A PENDULUM

A speed of change of angular displacement which always tends in a direction so as to oppose the angular displacement, which always adjusts to oppose the direction of the speed of change of angular displacement.

Figure 4 is the causal loop diagram which the group developed from the above model.

The startling results of these analyses are twofold. First, the similarity of appearance of the two system models was remarked upon. Secondly, the even more strongly counter-intuitive feedback mechanism than that discovered to operate in the case of a fire discovered to lie behind the behaviour of a pendulum was the source of much argument, discussion, debate and disagreement—even among the experienced physicists in the group. Intuition dictates that a pendulum must be a goal-seeking system,

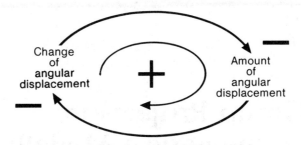

Fig. 4. Causal loop model of 'a pendulum'.

that negative feedback must operate here. Yet clearly the system is driven, counter-intuitively, by a positive, or growth promoting, feedback loop. The reader is left to study the results of the modelling experiment for himself and to convince himself of the validity, or otherwise, of the models developed.

In conclusion, past discussions with members of a more orthodox school of modelling approach make it imperative to add that the above described segment of an approach to model conceptualisation, formulation and construction is not to be viewed as a competing methodology, to be selected in preference to this or that particular approach to model abstraction. Instead the above should be used whenever possible to enhance discussions and encourage debate and deepen the understanding of any system that it is the intention of the modeller to model. The method should be employed to broaden skills participation through the involvement of a wider audience on account of its ease of appreciation by the non-modelling fraternity. The method primarily uses words to explore the major relationships perceived by the modeller to be responsible for the behavioural characteristics of the system under study and though the method lends itself more to those areas of study which are not the traditional hard sciences it is the conviction of the author that nothing but a more profound feeling and understanding for many 'hard' systems devolves to the scientist who is prepared to augment his approaches to model construction by consideration of the above points.

7
Group Projects in Mathematical Modelling

G. L. Slater
Sheffield City Polytechnic, UK

SUMMARY

Questions which often face us when trying to integrate a modelling component into a mathematics course are the following. How much modelling should we expect students to do? What are they likely to learn? Should we allow them to work in groups? What are we trying to assess and how should we do it?

This chapter discusses these questions in the light of one year's set of projects from the HND in Mathematics, Statistics and Computing at Sheffield City Polytechnic.

1. INTRODUCTION

In many institutions, a modelling project is part of the curriculum for students on degree or HND courses involving mathematics. At Sheffield City Polytechnic, one instance of this is in Part II of the sandwich HND course in Mathematics, Statistics and Computing. As in many such courses, the students work in groups of two or three, each group being supervised by one member of staff. This chapter is concerned with looking at the set of these projects for the academic year 1984–85, and making some assessment as to the worth of this part of the course.

2. THE COURSE

The Higher National Diploma in Mathematics, Statistics and Computing course has an entry qualification of one A-level in mathematics, the majority of students having a D or E grade pass. Their first acquaintance with the ideas of modelling comes as a brief introduction in the first year, as part of a mathematical methods course. However, this is little more than an introduction to the modelling loop (see, for example, Mason, 1984) and the presentation of some well-tried models. The modelling course is scheduled for the first term of the third year, immediately after the one-year industrial placement. The course is timetabled for six hours/week in two three-hour blocks. Somewhat unfortunately, these are Monday morning and Friday afternoon—not times expected to be conducive to fruitful modelling.

The first six weeks of the course are devoted to practical modelling, with the students finding out by doing. At the beginning of the term, the sessions are mostly on problems to which a fairly short first itcration around the modelling loop seems likely, so that the idea of performing a second iteration is acquired early in the course. Good students, of course, rapidly get into the problems and wish to improve their model several times. The problems at the beginning of the session are not expected to involve difficult mathematics; one of the disadvantages of running a sandwich HND is that little academic work is remembered after a year on placement in the real world. As the term progresses, skills increase and mathematical ability returns; the problems also grow in complexity, and students are encouraged to provide a more sophisticated approach. The schedule includes some old favourites, most of which have been published, which certainly do not all originate from Sheffield City Polytechnic. Many of these papers can be found in journals such as the *Teaching of Mathematics and its Applications* and references to most of the topics are given at the end of this chapter.

Week	Monday	Friday
1		Introduction
		Rugby goalkicking
2	Random response I	Random response II
3	Village school	Technological innovations
4	Sand quarrying I	Sand quarrying II
5	Barnicle geese	Introduction to extended modelling exercises
		External speaker
6	Oscillating bar I	Oscillating bar II

The next four and a half weeks of the course is spent with the students working on their projects, known officially as extended modelling exercises. Students see their supervisors roughly once a week whilst working on their projects; the term culminates in presentations to the whole class, by each project team, in the presence of most of the project supervisors and of the BTEC External Moderator. At the end of the presentation session, the students must submit their written project.

3. THE PROJECTS

Most of the project titles are chosen by the staff from their own interests; some of them are original, but mostly projects arise from articles appearing in journals. About 20% of the projects are duplicated from the previous year, but usually with different supervisors and hence a different emphasis. The rest of the projects are new to the course. This is not an organised policy but, having supervised a particular project once, staff seem reluctant to face the same problem again!

The students choose their own teams in which to work for the projects; teams were prescribed to have two or three members. It is interesting to note that, as usual, the girls group together as do the eccentrics. More surprisingly, weak students seem to stay together; the desire to hide their deficiencies from their colleagues being stronger than the desire to get better marks by working with a better student.

Students select their project after a five minute presentation by the staff member who is offering the project. More projects than student teams are offered, and the students vote for their choice of project. Usually, everyone finishes up with their first or second choice, but sometimes backstage negotiations must be carried out! It seems that students do not vote entirely on the basis of the topic offered for the project. Other factors, such as the personality and/or familiarity of the supervisor and the involvement of microcomputing in the project, seem to be significant.

The projects undertaken in the year were as follows.

Snooker	Establishing the likelihood of potting a ball and analysing one particular shot
Ordnance survey	Validating mileage claims by modelling distance measuring on a map
Secret messages	A survey of methods of sending secret messages and a detailed investigation of the RSA method
Stats packages on the BBC	An investigation and attemp to write one
Parent-teacher evening	Designing and implementing an algorithm to timetable a school parent-teacher evening
Shuddersfield	The siting of a new shopping centre
Computer configurations	An evaluation study to gather information on our mainframe system and answer certain questions related to performance
Cistern analysis	Modelling the flow of water into a cistern
Shot putt	What is the optimal angle for putting the shot?
Tape counting	To model the counter reading on a cassette recorder and the time elapsed in playing a cassette tape on the recorder
Advertising and sales	An investigation into the way in which sales figures are affected by advertisement
Broken glass	A problem involved in the occasional breakage of a float glass sheet and its prevention by the appropriate positioning of rollers

Industrial location	Minimising costs in siting a warehouse given different customer location patterns
Athletic performance	Examines the performance of male and female athletes at various running events

4. THE SUPERVISORS

As in most mathematics departments, the staff in our department can be classified into two disjoint categories—modellers and non-modellers. Modellers have some idea of what modelling is all about and have therefore carried out some modelling themselves. Non-modellers may well have heard all about modelling at great length but, because they have not actually tried any for real, they do not understand the difficulties experienced when modelling or, indeed, the distinction between models and modelling. A different categorisation of the staff can also be made into computer-literate and computer-illiterate. These two categorisations obviously have an effect on any project supervised. A wide range of staff becomes involved in this project exercise, mostly by volunteering; however, those who are press-ganged are usually offered a project that has been run previously to supervise. Although this way seem kind, it does not give the members of staff the experience of actually trying to do the modelling themselves, which would be valuable experience for non-modellers and modellers alike.

What staff expect of the projects and of the students also seems to vary tremendously. Some staff expect to have to show the students how to tackle the problem and, of course, they have to. Some expect the students to solve the problem all alone and, of course, in most cases they do not. Some expect their students to report progress every week and they do, some expect the students to see them only when they are in difficulties—this seems to have mixed effects. Some staff concentrate on the presentation of the final report, and expect 50 page word processed, beautifully presented, bound reports and usually get it. Others expect a few sheets of handwritten notes and, of course, they get it. Some staff expect well presented talks and help by giving practice runs with advice and help in presentation skills, others give no help with the oral report, usually with the obvious consequences. These differences are particularly obvious in the preparation of OHP slides for use in the talks. It might be reasonable to assume that students who spent a significant portion of the week looking at OHP slides would know what went into a good one, and what was and what was not visible, but it does not seem to be so.

Supervisors also have different approaches to arranging meetings with their students; since these sessions are not timetabled, finding time is a problem for all, but we see a lot of different solutions to this problem with meetings at 8.30 a.m. or after 5 p.m, or in the pub at lunchtime. How important is a modelling project?

5. THE OUTCOME

What, then, was the outcome of these projects? What exactly is modelling is a difficult judgement to make but, in my estimation only 11 of these 14 projects show any modelling at all. This may not be as desperate as it sounds; some of the projects were never really intended to be modelling exercises, they were interesting project topics and were presented to the students as such. The classification into projects containing modelling and their supervision by modellers or non-modellers is interesting.

	Supervisors	
Projects	Modellers	Non-modellers
Modelling	9	2
No modelling	2	1

If we consider how many of the projects actually closed the modelling loop, comparing their results with reality and at least suggesting ways in which their model might be improved the results are even more interesting.

Projects	Supervisors	
modelling	Modellers	Non-modellers
Loop closed	7	0
No loop	2	2

An investigation into the level of mathematics used in the 14 projects show that six were actually using the mathematics learned on the HND course—however, the title Mathematical Modelling is something of a misnomer: five involved arithmetic only, three involved O-level mathematics only, six involved HND 1 level mathematics.

The computing used can be classified by none, package use, actual coding and also by whether or not the supervisor is computer literate. Looking at these in the light of which project actually involved some modelling, we can see that few projects had time for modelling and coding.

Project	Supervisor	
computing use	Computer literate	Non-literate
None	4	3
Package	2	0
Coding	4	1

Project computing use	Modelling	Non-modelling
None	7	0
Package	2	0
Coding	2	3

6. PRACTICAL CONSIDERATIONS

The assessment of these projects is clearly a difficult procedure. There are the oral and written elements to balance, and problems in measuring and allowing for the amount of help given to the students and for the expectation of performance made. The scheme we use is as follows.

Talk

Clear statement of problem	5
Discussion of model	
(1) Assumptions	5
(2) Outline of method of solution/analysis	5
(3) Presentation of results	5
Discussion of results	5

Report

General layout	
(1) Division into appropriate sections	10
(2) Neatness	10
(3) Use of English	10

Content

(1) Introduction to and clear statement of problem	10
(2) Discussion of assumptions	15
(3) Error-free analysis	15
(4) Clear presentation of results	15
(5) Discussion of results	15

If the report contains none of supervisor's work, we multiply the mark obtained above by 1.0. If the project has been substantially directed by the supervisor, with a large contribution to the modelling and the analysis, we multiply the mark by 0.7. Otherwise, we allocate a multiplying factor in the range [0.7, 1.0].

Oral presentation is marked by all the staff present, which is usually somewhere between 5 and 10 in number. The report is marked by the supervisor; we try to double mark, but this is not always possible. The difficulty of one of the students doing most of the work is always present. In these circumstances marks are allocated in proportion to the amount of effort put in, usually after discussion with the individuals. Investigating the marks shows that projects either score well on the oral or on the report, but rarely on both.

	Report	
Talk	>65%	<65%
>60%	1	5
>60%	5	3

It is, however, worth noting that the range of marks allocated to the projects was 40% to 78%. Are all our projects average to just above average?

On a different practical level, how efficient are these projects in terms of teaching resource? Staff supervising a project are allowed 0.5 hour per week for the duration of the project. With 29 students on the course and a student timetable of 20 hours, the projects run at an SSR of 17.4:1. This cannot be described as extravagant, even after adding course overheads.

7. WHO LEARNED WHAT?

It may be a little late, but it is worth now considering the avowed aims of this project.

In this part of the course students gain experience in the activities of mathematical modelling—working in small groups, sharing tasks within the group, taking responsibility for a section of a project, presenting oral and written reports.

It cannot be argued that these aims are not met. However, what is it that we are really hoping for? Students certainly learn to work together in groups, they learn to work to a very tight and fixed timescale, they try their hands at oral and written communication in a more meaningful way, and they also learn to defend their work in discussion. Subsidiary benefits are practice in acquiring word processing skills, and increased use of micros. The experience of starting a piece of work not knowing which of their mathematical skills they will need, the necessity of trying out their ideas and possibly exposing them to the ridicule of their fellow students and, of course, in many cases, experience in data collection, are all valuable. Most of the students do seem to learn most of these things.

For staff, the involvement of a large proportion of their number in the same course is valuable, and in most cases the relationship between staff and students improves. For those staff press-ganged into assistance, there is probably some experience of modelling, and certainly the exposure to new teaching methods. Some are even forced to come to terms with new technology by their students' enthusiasm.

8. WHERE TO FROM HERE?

We firmly believe that our modelling projects are worthwhile as they stand. However, the course is being revised, and so changes are coming. We are keeping the projects and extending their ethos into a new area, that of workshops. The idea of the workshops is to provide a forum in which a multitude of activities, which can broadly be described as interdisciplinary, may be developed. The workshops will provide a laboratory environment for the learning of mathematical sciences, and a focus for both students and staff on cross-disciplinary applications work. These workshops will start with a pilot in the next academic year. Currently they are involving a great deal of staff effort in the preparation of courseware, particularly micro software. I expect there will be a report on the progress of the workshops at the next International Conference on the Teaching of Mathematical Modelling!

Some of the aims of the projects may well be achieved by the workshops and it is hoped that in this way our projects may be able to develop a greater modelling flavour than at present.

REFERENCES

Battye, A. R., Jukes, K. A. J. & Stone, J. A. R. (1984). The oscillating bar: a simple model in dynamics, *Teaching of Mathematics and its Applications*, 3, 30–36.

Callender, J. T., Slater, G. L. & Stone, J. A. R. (1984). Yet more mathematics of voting systems, *Teaching of Mathematics and its Applications*, 3, 18–23.

Callender, J. T., Slater, G. L. & Stone, J. A. R. (1985). The Random response model: a second attempt, *Teaching of Mathematics and its applications*, 3, 113–114.

Challis, N. V., Gretton, H. W. & Bradshaw, T. J. A simple water modelling case study, *Teaching of Mathematics and its Applications*

Green, D. R. (1979). Place the school, *Journal of Mathematical Modelling for Teachers*, 2, 1–6.

Huntley, I. D. & Ingledew, R. (1982). Rugby goalkicking, Solving Real Problems II, J. S. Berry, D. N. Burghes & I. D. Huntley (eds). CIT Press.

Mason, J. (1984). *Teaching and Applying Mathematical Modelling*, J. S. Berry, D. N. Burghes, I. D. Huntley, D. J. G. James & A. O. Moscardini (eds). Ellis Horwood.

Open University, Barnacle geese, TM281 Course.

Walker, L. A. (1986). Sand Quarrying, *Case Studies in Mathematical Modelling*, I. D. Huntley & D. J. G. James (eds). OUP.

8

Mathematical Models in the Teaching of Systems Thinking

E. F. Wolstenholme
University of Bradford Management Centre, UK

SUMMARY

This chapter reports on the experiences gained over the past few years in developing and using mathematically based models to introduce students to systemic concepts at Masters level in Business Administration at Bradford University Management Centre within a management science framework.

The chapter examines both the perceived need for management courses which promote systemic thinking and the background rationale leading to the development of the methodology used. This methodology is referred to as Qualitative System Dynamics (QSD) and provides a set of procedures for system description and analysis using influence diagrams which, although qualitative, gains structure and rigour from its mathematical origins. The approach is presented, a description given of the way it is taught, and comments made on the experiences gained to date from the teaching.

1. INTRODUCTION

One of the greatest challenges facing analysts involved in current management practice is how to provide managers with significant insights into complex problem situations with minimum resource input; particularly that of time. Whilst this is an old dilemma, it is increasingly taking on a new perspective as the complexity of modern management increases. First, it is becoming more and more apparent that conventional mathematical model-building practice often proves too slow, too restrictive and too inflexible in coping with the mixture of hard and soft issues, inherent in such problems. Further, the value added by the increased level of sophistication generated by a sizeable modelling effort is often questionable. Secondly, there is also an increasing awareness of the dangers associated with some of the current methods aimed at overcoming these problems. In particular, the increased use of techniques arising out of predefined problem types (as characterised by operational research) tends to lead to a blinkered, detailed and reductionist approach to problem analysis.

The premise of this chapter is that there is a growing need to develop general frameworks and methodologies rather than techniques, which can be taught to both managers and analysts and used by them to develop systemic overviews of problem areas. The objectives of such frameworks should be to provide a basis for structuring of concerns, to aid problem exploration, to improve communication, and to act as a focus for transferring modelling and life experiences to new problem areas.

This chapter is concerned with describing the development of a methodological approach for problem analysis which attempts to meet the above objectives. The method is based on system dynamics and arose out of experiences in teaching and applying this along with other systems methods and operational research in both an academic and industrial environment.

The origins of the method will be presented together with an outline of the procedures used and a description of its incorporation into courses at the Masters level in Business Administration at Bradford Management Centre.

2. SYSTEM DYNAMICS, SYSTEM ENQUIRY AND SYSTEMS METHODOLOGIES

The conception and early development of system dynamics took place during the late 1950s at the Massachusetts Institute of Technology (Forrester, 1958), and although early work was in the management field (Forrester, 1961) the subject became primarily known during the late 1960s for its application at the macro level in urban and global modelling (Forrester, 1971; Meadows, 1972). Although macro applications are still in evidence (Forrester et al., 1984), the scale of application has generally reduced and diversified during the 1970s. More recent work has seen an expansion in the development of its subtechniques and philosophy (Randers, 1979; Coyle, 1979; Wolstenholme, 1982; Richardson and Pugh, 1983).

The method as currently practised is still, however, essentially a socio–economic modelling technique concerned with the development of mainly large scale simulation models for policy analysis.

It is undoubtedly a powerful tool for providing insight into the behaviour and evolution of complex systems, which is somewhat unique in encompassing the adaptive nature of feedback effects of such systems. The procedure employed primarily represents a direct search for a model structure capable of providing a feedback hypothesis for observed system behaviour, and its ultimate objective is to move as quickly as possible to designing system policies by which to improve behaviour. Whilst this is a flexible and commendable procedure which can cope with subjective relationships it does, nevertheless, still suffer from the problem of being time consuming and restrictively carried out in a totally quantitative mode. In the opinon of this author, however, it is considered to provide an underlying richness of concept which could be applied independently of computer based simulation. There are already some steps being taken within system dynamics along these lines, but these do not go far enough. For example, there is already much work being undertaken in the qualitative use of influence and causal loop diagrams (Roberts *et al*., 1983) for explaining system behaviour. However, this tends to be carried out within a self-contained framework and it is felt that there is a need to present it within a more open and accessible framework.

Such frameworks exist in many fields and are created to facilitate stepwise procedures for structuring complex issues. These frameworks are usually qualitative since they must appeal to the full breadth of the fields involved, but they often contain optional quantitative components. A good example of such frameworks are those used in business policy (Gluek, 1976; Porter, 1980) for strategic assessment of companies. Here, the subsumed quantitative component is usually a hard technique area for assessing alternative strategies, after their generation, and before the implementation stage. The use of system dynamics as an alternative to existing frameworks in the business policy field is currently being explored (Morecroft and Paich, 1985).

The ultimate extension of the framework concept leads to the idea of the possibility of creating a general methodology for system enquiry, applicable across all fields. The search for such a general methodology for system enquiry has, in fact, been in existence for some time, with very limited success. It is the premise of this chapter that it is at this totally general level of enquiry that the concept of system dynamics as a framework for analysis should be aimed. It is suggested that the procedure of system dynamics provides an ideal basis around which to create a general system enquiry methodology.

Systems enquiry is used here to define the whole field of investigation concerning the understanding and design of change in complex human activity systems. This field is extremely large and although much attention has been increasingly focused on it in recent years, there remains a dearth of methods available to provide frameworks for anlaysis.

A belief in the need for holistic thinking has existed for a very long time, and its advantages over reductionist attitudes has been well expounded (Popper, 1957; Bertalanffy, 1968; Churchman, 1968). However, the development of meaningful methods by which to apply holistic ideas has so far proved very difficult, certainly in any practical rather than theoretical sense, although the literature is well sprinkled with attempts. These attempts come from a wide variety of disciplines. Discounting for a moment the methods of system dynamics, there are those arising out of the isomorphic elements of systems theory (Jenkins, 1969; Churchman, 1971; Hall, 1982), those resulting from attempts to expand and elevate mathematical problem-based techniques (Ackoff, 1972), those concerned with the wider interpretations of cybernetics (Beer, 1972), those based on the method of computer systems analysis (De Neufville and Stafford, 1980), those based on highly sophisticated structural modelling ideas (Linstone, 1978), and those based on purely qualitative diagrammatic and verbal procedures (Checkland, 1982).

The difficulties in generating useful methods centre on the compromise required between the vagueness necessary to be sufficiently general and the precision needed to produce specific results. In terms of problem analysis this dilemma takes the form of a need to have a wide and flexible approach to facilitate structuring of symptoms and problem identification whilst simultaneously requiring a narrow rigid approach to facilitate the creation and testing of remedies.

Consequently, there continues to be extensive research into compromise approaches for system enquiry, based on a mixture of hard result-orientated techniques and soft subjective methods, and current systems work is characterised by the search for improved methodologies. Methodology is defined here as the overall process of investigation, usually stepwise and iterative, by which concepts philosophies and theories can be expressed independently of the subject matter of the investigation and independently of the problem type to be considered. This use of the word methodology is not to be confused with its use in a specific technique sense, where it simply implies a list of steps necessary for the application of that technique, for example the linear programming methodology. The ideal methodology, according to Checkland, must avoid the content-free methodologies derived from general systems theory and the over-precise goal-orientated formulation stemming from system analysis.

3. A SYSTEM METHODOLOGY BASED ON SYSTEM DYNAMICS

The credentials of system dynamics as a system methodology have been explored elsewhere (Wolstenholme, 1982; Wolstenholme and Coyle, 1983), and a stepwise procedure put forward for system description and problem exploration. The basic points which were made concerning the credentials of the system dynamics method as a systems methodology were twofold. First, that the building blocks of rates and levels provided an excellent compromise between generality and usefulness for structuring systems, and secondly that the concept of control and its effect on system evolution over time was addressed. On the limitations side it was, first, suggested that there was a need to overcome the almost total emphasis placed on system processes in system dynamics, and that there exists a great need to recognise the role of the organisational structure of systems in determining system performance. This need is, in fact, becoming more widely recognised (Morecroft, 1984). Secondly, that there was a strong case for divorcing the system description and computer analysis phases of system dynamics.

It was with these ideas in mind that the following definition of a system dynamics as a two-part systems methodology was suggested.

A rigorous method for system description, problem exploration and analysis of change in complex systems, which facilitates and can lead to quantitative modelling and dynamic analysis for the design of system structure and control.

The first part of this definition is referred to here as 'Qualitative System Dynamics' (QSD) and the second part as 'Dynamic Simulation Analysis'. It must also be noted that QSD itself is split into two stages. First system description and problem exploration involving the creation of diagrammatic models and, secondly, qualitative analysis involving the analysis of the diagrams. A stepwise procedure for stage 1 of QSD is presented in Appendix A. This procedure essentially restates the process of model conceptualisation used in system dynamics; however, it places more emphasis on the early recognition of system levels and orientates the process towards a more conventional methodological form. The purpose of the procedure is to focus on influence diagramming for system description as a stand alone method, and to contrast its attributes relative to many of the less rigorous methods currently used in the system field. In the context of the whole spectrum of system enquiry, these description methods range from poetry at one end to totally explicit computer-based algorithms at the other. The use of influence diagrams, which are almost independent of system behaviour considerations, is not totally unique in the system enquiry field (Eden et al., 1979). The main characterisation of the procedure is that it starts at a high level of aggregation, and is aimed at facilitating both the elimination and introduction of resource states, as well as attempting to focus on the right level of resolution for the diagram. The overall objective of the procedure is to produce the simplest diagram structure, capable of relating the key variables associated with the cause for concern specified. The procedure is close to that recommended by a number of system dynamics practitioners for use in initial problem structuring, where both knowledge of the system is poor and a basic feedback hypothesis does not exist. In other words its final objective is to uncover any feedback loop structure.

The second stage of the procedure relates to the qualitative analysis of the derived diagrams. The definition of the steps for qualitative analysis suggested here may be viewed essentially as an attempt to replicate the basic procedures of system dynamics policy design, but without resorting to computer simulation. The important differences are that more attention is paid to the organisational implications of the system, and that the whole process of the analysis is slowed down and the degree of resolution of the analysis is increased. This is aimed at increasing communication and understanding of the system to assist the role of the system actors and owners in designing and implementing change for themselves.

A four-phase procedure is suggested for qualitative analysis and these phases together with the main steps involved in each are summarised in Appendix B. The major phases consist of static analysis, the identification of control issues, the dynamic implications of the existing system structure and the identification of factors for improving system performance. Many of the steps are self-explicit and are laid out to facilitate the application of the methodology by people who are not familiar with system dynamics. The details of each step will not be repeated here in the text, but discussion will be made of the general structuring of the procedure.

The first three phases are designed to make analysts stop and think about the system as is, rather than jump headlong into system redesign. The first of these stages focuses on creating a static feel for the process/organisational balance of the system and is geared to focusing the system actors' attention on the process perspective. The second tries to examine the existence or otherwise of control in the system and to categorise control, where present, by its variables, mechanisms and frequency of application. Only during the third phase is it suggested that true dynamic analysis and hand simulation of the structure be attempted.

The fourth and final phase of the procedure is concerned with changing the system. It is at this point that we are faced with the problem of how, in the absence of a quantitative test procedure, objectively to improve system performance and to overcome the original cause(s) for concern associated with a system. This problem is, of course, not unique to this particular approach and is a common difficulty encountered by all soft system methodologies. In fact, by having used system dynamics as a basis for structuring the system it could be argued that there exists a much stronger basis for analysis than in the case of many other system methodologies and that this is where system dynamics scores an enormous advantage over other methods. There are two reasons for this, both associated with the relatively high level of rigour attached to the diagramming procedure, which has arisen out of its simulation origins. First, the level of communication facilitated by the diagrams is high and they are very orientated towards encouraging self-diagnoses and self-help amongst the system actors and owners. Secondly, there exists a whole body of proven general results for system dynamics structures which can be used as a basis for directing change in systems. Whilst it is not suggested that such results will have a relevance in all systems and that there are, undoubtedly, dangers in the indiscriminate use of them, it can be argued that there is, in the majority of practical systems, sufficient scope for improvement to justify their use. This idea of the identification and transfer of generic components and results between dissimilar systems (isomorphism) is one of the most deeply established concepts associated with the development of system methodologies (Von Bertalanffy, 1968) and is an area being strongly pursued at present in the system dynamics field (Morecroft and Paich, 1985).

It is no doubt, likely that all system dynamics practitioners would, if asked, be able to produce a good and useful breakdown of generalised results that could usefully be transferred between systems. The list identified in phase IV of the methodology here is neither claimed to be sufficient nor definitive as it is still under development. The list presented concentrates on the fundamentals of improving any mismatch between the organisation structure and process structure of the system, in highlighting the need for control, in defining how control may be designed or improved in terms of objectives and discrepancies, and by reducing delays in information and hence monitoring needs of the system. It is further suggested that a search is made within the diagrams developed for subconscious feedback. That is, longer term feedback loops which often exist between system variables, but are not directly perceived as important by system actors because they perhaps exist at a different level of aggregation or on a different time-scale from the loops under scrutiny. Finally, it is suggested reiteration takes place, and that the dynamic implications of any new structures defined are examined as in phase III of the procedure.

4. TEACHING THE METHODOLOGY

Two courses have been developed based around the two parts of the methodology defined and these are each presented in eight-week modules of two hours per week to students at the Masters level in Business Administration. The first course is entitled 'Systems Methods for Strategic Thinking' and is centred on the use of QSD. The second is entitled 'Dynamic Simulation Modelling' and is presented to enable the more quantitative students to develop simulation models for the diagrammatic models created in the first course. Outline syllabuses for these courses are presented in Appendices C and D respectively. Since the latter course is of a more conventional nature, emphasis here will be placed on a description of the former.

The content of the first two weeks of the course centres on reviewing the need for systems methods in management, describing the origins and general philosophy of holism and the difficulties of creating methods capable of applying holistic concepts to management in any meaningful and acceptable way. Methods currently in use for system diagramming, system classification and for defining system boundaries are explained. The diagrams presented here range from those of the non-conventional one-off type, through Venn diagrams, to flow block diagrams and systems maps and on to algorithmic diagrams. The concept of systems methodologies are introduced at this stage and an exercise undertaken using one such methodology.

In weeks three and four the methodology of QSD is introduced and exercises used which build up to providing experience in its ultimate use; that is, in providing a means for analysts and managers to create initial inroads into new problem areas. To achieve this, exercises are given in three degrees of specificness. The first one requires influence diagrams to be drawn and analysed based on comprehensive and unambiguous written descriptions of system variables. The second requires an influence diagramming structure to be created from typical, slightly ambiguous written descriptions of problem situations, but with some help given in the identification of the most important system states to be represented.

The third type of situation presented requires that a diagrammatic QSD model be created and analysed from a typically vague remit of the type encountered in practice, where senior managers express a general cause for concern and require guidance in clearing their thinking, rather than definitive answers. It is this type of situation that current management training pays little attention to, the belief being that if the analysts have their array of techniques well understood, then they will cope. In fact, in both the real world and in the classroom situation, experience shows that this is the type of situation which analysts are least able to cope with and one where panic most easily sets in. The main reason for such feelings is that original thinking is demanded and this is not an attribute that too many people can call upon without help. Even with help from methods such as the one described here, most students initially still rely entirely on instinct and experience. It is generally found to take some time before the taught framework is established as a catalyst for instinctive thinking. Once this has happened progress in problem analysis is greatly improved.

Obviously analysis is still more effective the better the knowledge the student has of the system in which the problem lies. Consequently, apart from case study presentations of QSD, a large part of the remainder of the course is taken up with encouraging students to apply the methods to systems within their own experience. Such exercises are invaluable because, ultimately, it is only when students demonstrate to themselves that new insights can be generated, into what were previously thought of as familiar and known systems, that the method is really proved for them. Taught methods are seldom put to this test as there is a risk involved, and consequently the credibility and total acceptance of the methods is never achieved.

Experience with the course here indicates that confidence in applying it does, in fact, depend on certain factors. The most important of these seems to be the degree of previous experience of management and work situations in general by the students. That is, appreciation is higher when they have seen at first hand the degree of 'mess' associated with management problems, and have experienced the difficulties of being objective and confident in identifying and carrying through relevant analysis. Acceptance of the method additionally depends on having an open mind, which perhaps simply means an intuitive appreciation of holism.

The true test of any methodology is, ultimately, that of whether it becomes a totally subconscious thing to do and affects thinking without awareness of this at the time of application. This is almost the case with most variants of the scientific method as practised in pure scientific research, and we are obviously a long way from this with the development of methodologies in other softer fields of enquiry. However, most students involved with the methods described here would say that there are two underlying features of the procedure which are most strongly retained and used at a subconscious level. The first of these is the system process perspective captured by the approach, which is often very new to many students. The second is the idea of the need to control process in an integrated way across organisational boundaries describing areas of responsibility. These factors are, in fact, the fundamentals of system dynamics originally perceived by Forrester, and the success of the methods in communicating these ideas goes a long way to eliminating a strong teaching concern which existed at the design stage of the courses. This was whether, in fact, it was possible to teach what was effectively a method of system structure arising out of computer simulation software, without recourse to the computer simulation software itself, and whether in fact an ability with influence diagram analysis could only arise out of previous experience with fully quantified simulation models. This issue has not proved to be a problematic factor at all, and the qualitative nature of the course has additionally assisted in increasing the scope of communication of ideas associated with process and control by encompassing non-management science students to undertake it.

The latter point is reinforced by the number of such students then interested in undertaking the second course, which introduces them via a fairly easy transition into the arena of simulation modelling. The main purpose of the second course is, in fact, to encourage quantative analysis of the problems developed and analysed qualitatively on the first course. It is hoped eventually, by comparing the initial qualitative analysis of these problems with the final quantitative analysis, to be able to quantify the magnitude of the value added from quantitative analysis.

5. CONCLUSIONS

This chapter has suggested that there is a need to expand the role of management science in the area of strategic problem analysis and has put forward an holistic methodology based on the concepts of system dynamics for this purpose. Experience with teaching this methodology to Masters students in Business Administration would suggest that there is much potential in the approach and that it forms a good compromise between precision and flexibility, in keeping with the current problem solving needs of managers who operate in an increasingly complex and subjective environment.

APPENDIX A: A STEPWISE METHODOLOGY FOR QSD (SYSTEM DESCRIPTION AND PROBLEM EXPLORATION)

(1) Recognise the key variables associated with the perceived cause(s) of concern in the system and with the remit provided for the enquiry. Where possible, examine the behaviour of these variables over time and define a time hrorizon for the analysis.
(2) Identify some of the initial system resources associated with the key variables.
(3) Identify some of the initial states (levels) of each resource using a level of aggregation compatible to the time horizon defined in (1).

(4) Construct physical flow modules associated with each state of each resource, containing the physical processes or rates which affect these. (A module must contain at least one resource state and one rate.)
(5) If more than one state of a resource is involved cascade flow modules together to produce a chain of resource conversion or transfer.
(6) For each module or set of cascaded modules identify the *intra* module behavioural information and control (policy) links by which the levels affect the rates.
(7) Identify similar behavioural and control links between modules of different resource types. For complex situations this should be carried out for small groups of resources at a time within a defined theme and the resultant diagrams reduced to produce the simplest possible, consistent with relating the key variables of the investigation.
(8) Identify any new states of existing resources, or new resources, which affect the rates of the modules created or new key variables, and add these to those recognised at (1) and (2) . Reiterate if necessary.

APPENDIX B: A STEPWISE METHODOLOGY FOR QSD (MODEL ANALYSIS)

Phase I Static analysis of the model structure
(1) Confirm with the system actors that the model relates the major system variables associated with the original cause for concern.
(2) Identify uncertain contentions or highly subjective relationships between defined variables.
(3) Group variables into sets characterised by existing areas of functional responsibility (such as common accountability) and superimpose Venn diagrams to delineate the boundaries of these.
(4) Identify delays: identify the order of magnitude of delays in both physical operations and in the retrieving or perceiving of information.

Phase II Identify control issues
(1) Search for control framework. Classify information links as behavioural or control based. (Behavioural links are defined as the means by which systems adapt themselves in the long term if left to their own devices, whereas control mechanisms are defined to represent the actions of humans aimed at changing system performance.)
(2) Classify resources by their control functions. If control links exist, identify the resource stream which is being controlled (the controlled resource) and the resource stream which is acting as a controller (the controlling resource).
(3) Identify the particular variables within the controlled resource, through which control is implemented and identify who is the controller (i.e. who has organisational responsibility) of each of these controlled variables.

(4) Clarify the mechanisms of control, i.e. identify the range of control policies for each controlled variable; identify the range of control policies for each controlled variable; identify the sources of information feeding the policies and the intermediate processes through these data pass; identify the mechanism by which the policies convert information into action.
(5) Determine the frequency of control implementation, that is, can control be instigated on a real-time continuous basis or only at certain review points? Is the frequency of implementation of control restricted by the speed of information retrieval or by organisational factors (e.g. committee meetings)?

Phase III Dynamic implications of the model structure
(1) Identify the major feedback loop structure of the model.
(2) For each feedback loop carry out a hand-simulation to assess its likely behaviour; first start by changing each of the controlled variables to extreme values of the policies defined for them and secondly by changing each of the uncontrolled (exogenous) variables, again to the extreme range of values likely to be experienced for them.
(3) Is there any evidence to suggest that the system will be subject to any well known counter-intuitive or self-regulating models of behaviour?

Phase IV Identify factors likely to lead to improved system performance
(1) Can the organisation structure be changed to better match the process structure, or vice versa? For example, could one person be given responsibility for more than one controlled variable in a particular resource stream? If this is not possible can further control be designed to help resolve conflicts?
(2) Do overall objectives exist for the whole or parts of the system defined and do these conflict?
(3) Can control be designed for variables that are presently uncontrolled or only subject to behavioural control?
(4) Does the concept of a desired state exist for each of the actual state variables in the system and are critical values defined for actual states? If so, are they themselves variables or constants? Does the concept of measuring discrepancies between actual and desired states exist?
(5) For each controlled variable within each resource flow is account being taken, in its control policy, of the content of upstream and downstream states of the resource.
(6) For each controlled variable are there any information flows that are very protracted and can these be short-circuited; that is can the system be made more responsive and is this desirable? It may be that attributes of the controlled resource could be monitored whilst it is within the controller's sphere of responsibility rather than outside.
(7) Examine the information retrieval and monitoring infrastructure of the system to make it compatible with the control requirements identified.

(8) Examine the likely links which might exist between variables in the system which are currently perceived as being totally independent.
(9) For each new defined policy, or change in system structure, created either intuitively or from the previous steps, repeat the process of defining its dynamic implications given in phase III.

APPENDIX C
Course 1: Systems methods for strategic thinking
Objective: To develop qualitative skills in interpreting and describing systems for the purpose of identifying, understanding and solving real world management problems.

(1) *The need for system methods*
An examination of the difficulties of current management techniques and practices for problem solving. A presentation of the philosophy of holism and its relevance to managers in developing an integrated viewpoint across functional boundaries. Introduction to methods for system description using diagrams.
(2) *Current methodologies for system enquiry*
An overview of practical methods for applying system ideas. Application of one such method.
(3) *The dynamics of systems*
Demonstration of the importance of examining systems in terms of their behaviour over time. The relationship between time behaviour of systems and their dominant operating policies. The control of systems via information feedback structure and policy.
(4) *Qualitative System Dynamics (QSD)*
Presentation of system dynamics as a method of system description and qualitative analysis.
(5) *The use of QSD in real life problem analysis (I)*
Presentation of small case studies using QSD.
(6) *Applications workshop*
Use of QSD in analysing students own or set problems.
(7) *The use of QSD in real life problem analysis (II)*
Presentation of case studies involving the use of QSD in highly complex, multiple ownership system.
(8) *Course Review*
Comparison of merits of the methods presented both relative to each other and relative to management needs and alternative approaches.

APPENDIX D

Course 2: Dynamic system modelling
Objective: To develop quantitative skills in system modelling for the purpose of designing system controls for improving system behaviour over time.

(1) *Principles of dynamic simulation I*
   A description of how continuous simulation software (DYSMAP) operates and the construction of basic level, rate and auxiliary equations from influence diagrams.
(2) *Modelling demonstration*
   Using the computer demonstration of the construction and running of a simple dynamic model to test alternative control options.
(3) *Workshop I*
   Construction of a company model in groups.
(4) *Principles of dynamic simulation II*
   A description of more advanced functions for dynamic modelling. Model validation.
   Workshop II
   Design of control experiments for the company model.
(5) *Principles of dynamic simulation III*
   A description of the use of dimensional analysis and the dynamics of delays.
   Workshop III
   Construction in groups of a set model of students' own choice.
(6) *Case Study I and Workshop IV*
   First presentation of a system dynamics application plus further work on model development.
(7) *Case Study II and Workshop V*
   Second presentation of a system dynamics application plus further work on model development.
(8) *Course review*
   Open discussion.

REFERENCES

Ackoff, R. L. (1972). *The Art of Problem Solving*. Wiley.
Beer, S. (1972). *Brain of the Firm: The Managerial Cybernetics of Organisation*. Allen Lane.
Checkland, (1982). *Systems Thinking Practice*. Wiley.
Churchman, C. W. (1968). *The Systems Approach*. Delta Books.
Coyle, R. G. (1979). *Management System Dynamics*. Wiley.
De Neufville, R. & Stafford, J. H. (1980). *System Analysis for Engineers and Management*. McGraw-Hill.
Eden, C., Jones, S. & Sims, D. (1979). *Thinking in Organisations*. Macmillan.
Few, W. (1981). System dynamics' silver anniversary: a time for reflection and reformation, *DYNAMICA*, 7, pt. II.
Forrester, J. W. *et al*. (1984). The United States National Economic Model.
Forrester, J. W. (1958). *Principles of Systems*. Cambridge, Massachusetts, USA: Wright-Allen Press.
Forrester, J. W. (1961). *Industrial Dynamics*. MIT Press.
Forrester, J. W. (1971). *World Dynamics*. Cambridge, Mass., USA: Wright-Allen Press.
Forrester, J. W. (1976). *Industrial Dynamics*. MIT Press.
Forrester, J. W. *et al*. (1984). The United States National Economic Model, *Proceedings of the 1984 International System Dynamics Conference, Oslo, Norway*.

Gluek, (1976). *Business Policy*. McGraw-Hill.
Hall, A. D. (1982). *A Methodology for System Engineering*. Van Nostrand.
Jenkins, G. M. (1969). The systems approach. *J. Systems Engng*, **1**.
Linstone, H. A. *et al*. (1978). The use of structural modelling techniques for technological assessment, *Technol. Forecasting Social Change*, **14**, 231–327.
Meadows, D. H. (1972). *Limits to Growth*.
Morecroft, J. (1984). System Dynamics and the Portrayal of Bounded Rationality, *System Dynamics Conference, Oslo*.
Morecroft, J. W. & Paich, M. (1985) System Dynamics for the Design of Business Policy and Strategy, Working paper from Alfred P. Slaon School of Management, 1606–84.
Popper, K. (1957). *The Poverty of Historicism*. Routledge & Kegan Paul.
Porter, M. E. (1980). *Competitive Strategy*. New York: The Free Press.
Randers, J. (ed.) (1979). *Elements of the System Dynamics Method*. MIT Press.
Richardson, G. P. & Pugh, A. L. (1983). *Introduction to System Dynamics Modelling*. MIT Press.
Roberts, N. *et al* (1983) *Introduction to Computer Simulation—A System Dynamics Modelling Argued*. Addison-Wesley.
Wolstenholme, E. F. (1982). System dynamics in perspective. *J. of the Oper. Res. Soc*. **33**, no. 6.
Wolstenholme, E. F. & Coyle, R. G. (1983). System dynamics as a rigorous procedure for system description and qualitative analysis. *J. of the Oper. Res. Soc*.

Section B
Models

9

Model Validation–Interactive Lecture Demonstrations in Newtonian Mechanics

C. R. Haines
The City University, London UK

and

D. Le Masurier
Brighton Polytechnic, UK

SUMMARY

It is evident that in attempting to model phenomena using Newtonian mechanics, many students encounter difficulties through their own lack of physical intuition. Some attempts can be made to alleviate the problem by setting up impressive demonstrations which make full use of modern TV and video techniques. Amongst others, the Open University in particular has met with considerable success in this area. However, these techniques are usually expensive to devise and to mount, and this acts as an inhibiting factor for many (smaller) institutes which have neither the equipment nor the technical support readily available to prepare, or even to show, them.

Such demonstrations are often remote from the students, and in this micro-age the initiative for innovative techniques in model validation is in danger of moving away from the lecturer. This chapter puts forward details of well-proven experiments and ideas for practical demonstrations which work, and which will be appreciated by students on GCE Advanced level courses as well as by undergraduates. A number of these experiments take a fresh look at very familiar problems, such as a ladder leaning against a wall and cylinders rolling down an inclined plane. They use simple models which may be constructed from common articles without the aid of a technician, and they seek to illustrate the main concepts of dynamics, such as force, momentum and energy. In many cases the students would be encouraged to take an active part in the demonstration thus becoming closely involved in working mathematics, and appreciate that this can be fun.

1. INTRODUCTION

Over the past few years there has been a fall in the number of students applying to read for a degree in physics and related subjects at universities and polytechnics. This is mainly because there has been a decline in the interest shown in these subjects by pupils in schools. This has also led to the admission of students to undergraduate courses in mathematics who have not been exposed to the modelling of physical phenomena using Newtonian mechanics. The exciting course developments in further and higher education, through the introduction of aspects of mathematical models and modelling at various stages in the undergraduate programme, have tended to concentrate upon such topics as population growth, traffic control and epidemiology. Of course, these areas are eminently suitable for inclusion in applied mathematics courses, but this has often been achieved at the expense of mechanics, which may have been neglected of late, since the examples used were always rather too stereotyped and unimaginative. This combination of circumstances has meant that, when it has been necessary to introduce students to concepts in mechanics, they have encountered difficulties through their own lack of physical intuition. An intuitive feel, in this sense, is acquired through experience and exposure to the right concepts and associated examples over an extended period.

Model validation is a vital feature of the mathematical modelling process, and whilst the topics mentioned above demonstrate newer uses of mathematics (Lighthill, 1978), it is evident that model validation can be impracticable. Even though expensive equipment may be provided, lecture demonstrations in real time are not always possible; for example, a time scale of weeks or months is common in many models in epidemiology. This chapter serves as a reminder that there are areas of Newtonian mechanics for which the model can be validated easily, directly, instantaneously and with some effect, without the necessary use of complicated apparatus.

One way in which ideas have been successfully reinforced is by illustrations which utilise modern TV and video techniques. The Open University, in particular, has met with considerable success in this area, and the increasing availability of VCRs indicates that video material does make a valuable contribution to the methodology of the teaching of mathematics. The demonstrations shown on TV and video are certainly impressive, but they are one-off experiments remote from the viewer, and therefore they are only partially successful as an aid to understanding. The presentation is often such that an element of contrived surprise is engendered, similar to that which is observed on the successful performance of a trick on a TV magic show. The latest developments have incorporated interactive features, through which the viewers can pace their learning by accompanying texts similar to those already available in audio tape learning systems (Open University, 1984). These TV and video packages depend upon considerable support both in terms of equipment and materials, and of technical staff.

The demonstrations in this chapter show that Newtonian mechanics is an area in which the live experiment can be more effective, and for which the expense of TV and video packages is unnecessary. The impact of the real time demonstration can be enhanced by monitoring the reaction of the students, and varying the topics covered and the manner in which they are presented to suit the particular audience in question.

The recent introduction of an interactive element into video packages is a natural development of a technique which is used to good effect in the computer simulation of physical phenomena. The use of readily available software packages on a microcomputer is seen by many educationalists as the latest in a sequence of ultimate teaching aids. Certainly, the MIME project at Loughborough (Bajpai *et al.*, 1984) lends support to this view, with packages developed which cover the whole of one GCE A-level Mechanics syllabus. The establishment of microcomputer laboratories for the teaching of mathematics enjoys the support of funding agencies in the UK, and there are interesting developments in this area, for example, at Birmingham University and Manchester Polytechnic. A wealth of experience has accumulated over the several years of operation of CATAM at Cambridge University, and the packages developed for the teaching of topics such as Fourier analysis show a very high level of expertise and are amongst the best so far available (Harding, 1984). The main feature of these computer-simulated experiments, and the associated software, is that the student can vary the parameters of the problem and can therefore learn in a structured way, through experiments. However, great care must be exercised in interpreting the results of computer modelling and validation persists as a problem. The exciting graphics qualities and the skill of the programmer invite the admiration of teacher and student alike, but give no feel for the problem and are often at the expense of a real understanding of the underlying concepts. In addition, despite its interactive nature, the total experience of computer simulation remains remote from the student.

This chaper proposes ways in which the initiative for innovative techniques can remain with the teacher. The real time lecture demonstration should retain its important role in the teaching programme, and should shrug off its old-fashioned image. The presentation and the skilful management of the Royal Society Christmas lectures is an instructive case in point. So, too, are the physics lectures of the Molecule Club at the Mermaid Theatre, and also the several set piece demonstrations rudely given at summer schools of the Open University. The demonstrations described in this chapter are a small sample of those that can be provided within the spirit of this Chapter. They should be given in an atmosphere in which the audience will appreciate that there are opportunities for things to go wrong and will eagerly anticipate seeing how the actors cope with disasters.

2. THE DEMONSTRATIONS

In making our choice we had in mind a number of criteria which must be satisfied. In discussing these at greater length we attribute no particular order of importance.

  (i) They must be simple and practicable, requiring only cheap, recognisable, and easily obtainable materials and equipment, and the minimum of technical support.
  (ii) Collectively the models must provide a good coverage of the basic principles of Newtonian mechanics. The desirability of model validation, and the benefits of *predicting* behaviour, should be apparent from the demonstrations.
  (iii) Depending upon the audience and the time available, there should be possibilities for variations and extensions.
  (iv) Opportunities for student interaction and participation must be provided. Whenever possible, they should be encouraged to experiment themselves. This approach to mathematics must be entertaining as well as instructive, and seen to be so.

We believe that all these conditions, at least to some extent, are met by the following five exercises:

(1) Slipping ladders.
(2) Beer can racing.
(3) Falling chains,
(4) Swinging oranges.
(5) Free falling oranges.

We now describe these in detail.

2.1 Slipping ladders

Apparatus: a short ladder, a smooth wall and a willing volunteer!

The applied mathematician, reared on a traditional diet of statics, can recall all the common examples of ladders leaning against walls, with various convenient coefficients of static friction to allow a deterministic solution of the problem. When we first decided to include this demonstration, we intended to show how the effect of friction could be modelled, and the part it plays in verifying Newton's first law.

Stage 1: The ladder is leant against a smooth wall, as shown in Fig. 1, so that it just stays in place. It is easy to measure the angle of inclination to the horizontal α, giving an estimate of the limiting coefficient of static friction, μ, between the floor and the ladder

$$\mu = \tfrac{1}{2} \cot \alpha$$

Stage 2: The ladder is now inclined at an angle θ ($\theta > \alpha$), Fig. 2, and the foolhardy volunteer is persuaded to climb it. Then we can predict when the ladder should start to slip.

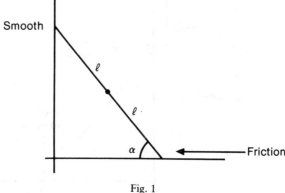

Fig. 1

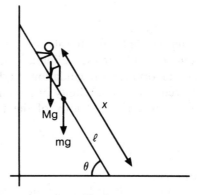

Fig. 2

If the volunteer has mass M and climbs a distance x up the ladder, of mass m, before it starts to slip then, resolving forces and taking moments appropriately yields

$$\frac{x}{l} = \frac{m}{M}\left(\frac{\tan\theta}{\tan\alpha} - 1\right) + \frac{\tan\theta}{\tan\alpha}$$

If, in stage 1, $\tan\alpha < 1$, then setting θ to $45°$ in stage 2 leads to $x > l$, so that, for a ladder of length $2l$, the unfortunate volunteer will get over half-way up before slipping occurs. Similar results hold for values of θ other than $45°$.

What happens if the floor is smooth and the wall is rough, as in Fig. 3?

An experiment should confirm the intuitive result that, since the only horizontal force acting is the normal reaction at the rough wall, then the foot of the ladder must move away from the wall, regardless of the magnitude of the frictional force at the wall.

In theory, these demonstrations could be said to be as easy as falling off a ladder! However, model validation is not always such a simple matter. The idealised smooth floor is not as common as one might expect, and our

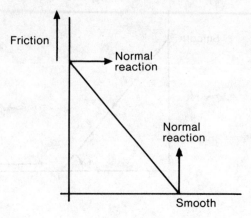

Fig. 3

experience suggests that smooth walls are even rarer, for we met with little success in our attempts at the validation of this particular model. This confirms the importance of validation, and the necessity of a critical appraisal of accuracy and reliability of a model. For the authors, it threw considerable doubt on the validity of the traditional model of limiting friction.

2.2 Beer can racing

Apparatus: A plank or board of minimum width 0.5 m, minimum length 1 m, assorted unopened beer or lager cans, of various sizes.

In these experiments we investigate the significant parameters of rotational motion of rigid bodies rolling down an inclined plane. We can show that the time taken for a uniform cylinder to roll down a given slope is independent of its size and mass, and depends only upon how that mass is distributed (i.e. its radius of gyration). The board acts as an inclined plane, and it must be sufficiently wide to accommodate two cans rolling simultaneously, in parallel, down the line of greatest slope. The plane is inclined by propping up one end, perhaps with some old video cassettes, to give an angle of approximately 20° to the horizontal. Of course, this may be varied to some advantage between, say, 10° and 45°. We rely on direct comparison to validate our models, thus avoiding the need for accurate measurements.

Figure 4 shows a uniform rigid body, such as a cylinder, sphere or hoop, rolling, without slipping, down a plane inclined to the horizontal at an angle α.

It can be shown (Chester, 1979), that the acceleration $\ddot{x}$ of the body down the slope is given by

$$\ddot{x} = \frac{a^2}{(a^2 + k^2)} g \sin \alpha$$

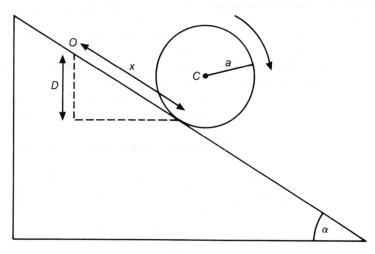

Fig. 4

and its speed $\dot{x}$, by

$$\dot{x}^2 = \frac{2a^2}{(a^2 + k^2)} g\, D,$$

where a is the radius of circular cross-section, k is the radius of gyration, about the axis of rotation, g is gravitational acceleration and D is the *vertical* distance fallen by the body, so that $D = x \sin \alpha$.

Noting that k is directly proportional to a, it is helpful to write

$$s(k) = \frac{2a^2}{(a^2 + k^2)}$$

so that $\dot{x}^2 = gD\, s(k)$, and to construct a table of values for $s(k)$ for the conventional rolling bodies.

Table 1

Body	$(k/a)^2$	$s(k)$
Solid sphere	2/5	10/7
Solid cylinder / Circular disc	1/2	4/3
Hollow sphere	2/3	6/5
Hollow cylinder / Circular hoop	1	1

The maximum values of $\dot{x}$ coincide with those of $s(k)$, and so we observe from Table 1 that the fastest rolling object should be a solid sphere, and the slowest a hollow cylinder or a circular hoop.

There are many possibilities available to the demonstrator, so that it is easy to adopt an approach suited to the audience. One favoured method is to persuade the volunteer, if sufficiently recovered from the activities on the ladder, to participate in a *slow* beer can race, namely to select one of two similar cans which will take *longer* to roll down the slope. You may allow a trial run, with two *unopened* cans, and then invite the gullible volunteer to choose a can, and to mark it accordingly. Now empty the other can and victory is assured since the empty can must be slower.

There are countless variations, particularly if you have a wide selection of rolling bodies. Before embarking on the experiments, the audience are invited to suggest the significant parameters, such as mass, diameter, and so on, and a careful sequence of trials should gradually eliminate most of them. As a further experiment planes, inclined at different angles, confirm that it is the parameter D that is significant for two identical bodies. We would also like to point out that experiments with sundry Guinness containers led to the conclusion that since $s(k) = 12/7$ for a *cube*, then a rolling cube should be the swiftest—this confirms the necessity for model validation! It is also possible to investigate the relevance of friction. However, unless the surfaces involved are exceptionally smooth, the speeds developed are rather high, and this can impair the quality of the beer.

2.3 Falling chains
Apparatus: One set of bathroom scales, one set of post-office-type scales with a large dial, and various lengths of heavy chain.

A layman may have an intuitive idea of momentum through experience of the fairground exhibitions of test your own strength, see Figure 5. He may not appreciate that as a mathematical concept, but he may be aware that as momentum is destroyed rather quickly, then an increased force results. This short demonstration aims to show that force, acceleration and momentum are related through Newton's third law.

Initially, it is desirable to persuade a volunteer to jump up and down on a set of bathroom scales, to demonstrate that this gives a heavy reading (and heavy breathing) compared with the static case. For the main

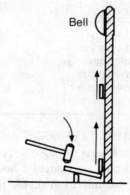

Fig. 5

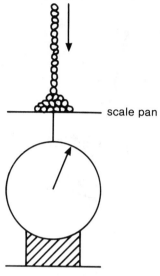

Fig. 6

demonstrations, the chains are weighed carefully on the post-office-type scales. The audience is then asked to estimate the dial reading when the chains are allowed to fall freely onto the scale pan, Fig. 6.

The theory (Ramsey, 1962), predicts that the maximum scale reading will be three times the total weight of the chain. The experiment is concluded by marking this particular point on the scale, and allowing the chain to fall again.

This simple experiment is very effective, and has been much appreciated by students (Thorpe, 1984). It uses the minimum of apparatus, which does not need fine tuning to make it work.

2.4 Swinging oranges

Apparatus: Two large oranges of equal mass, length of string (approximately 2 m), skewer or small screwdriver, paper clips, metre rule.

The authors have come across a number of interesting examples of normal mode motion (Fendrich, 1982), but the one described below, based on an idea gleaned at Brighton Polytechnic (Bell, 1977), has the great merit of simplicity. It validates the natural frequencies and corresponding normal modes of a model of a double pendulum. This is prepared by threading the string through the core of each orange, using the skewer.

Ensure that one orange, o_1, is fixed at the mid-point of the string, and the second orange, o_2, at one end. Paper clips provide a useful means of fixing. The pendulum so formed is suspended from a fixed point A, so that it can swing freely in a vertical plane, see Fig. 7. If necessary, the weary volunteer may be coaxed to stand on a desk to act as a fixed support. In practice, several members of the audience may be encouraged, with a little guidance, to undertake the complete experiment.

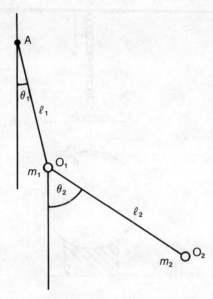

Fig. 7

A full treatment of the theory, using Lagrange's equations, appears in Chester (1979). A more elementary, yet more tedious method can be used to derive the following equations of motion, which have been linearised on the basis of small oscillations.

$$(m_1 + m_2)l_1^2\ddot{\theta}_1 + m_2 l_1 l_2 \ddot{\theta}_2 = (m_1 + m_2)gl_1\theta_1$$
$$m_2 l_1 l_2 \ddot{\theta}_1 + m_2 l_2^2 \ddot{\theta}_2 = -m_2 g l_2 \theta_2$$

For our problem, we have

$$m_1 = m_2 \quad \text{and} \quad l_1 = l_2 = l, \text{ say,}$$

so that, in matrix form, the equations become

$$\begin{bmatrix} \ddot{\theta}_1 \\ \ddot{\theta}_2 \end{bmatrix} = \frac{g}{l} \begin{bmatrix} -2 & -1 \\ -2 & 2 \end{bmatrix} \begin{bmatrix} \theta_1 \\ \theta_2 \end{bmatrix}$$

so that the natural frequences n_1, n_2 of the system are given by

$$n_1^2 = (2 - \sqrt{2})\frac{g}{l} \quad \text{and} \quad n_2^2 = (2 + \sqrt{2})\frac{g}{l}$$

with corresponding normal modes

$$[1 \ \sqrt{2}]^T \quad \text{and} \quad [1 \ -\sqrt{2}]^T.$$

First, random motion is illustrated by releasing each orange from a convenient point distinct from the position of static equilibrium (Fig. 8).

It is important to keep the string taut. Release from rest should yield a comparatively complex motion, to start with. Next release o_1 from a point

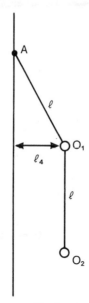

Fig. 8

at a horizontal distance of 0.25l, and o_2 0.6l, from the vertical through A, as shown in Fig. 9.

The oranges should swing in phase, with approximate simple harmonic motion. Repeat the process, asking a student to note the time, say T_1, of five complete oscillations. Now release o_1 from a point at a distance 0.25l to the right, and o_2 0.35l to the left, relative to the vertical through A, as in Fig. 10.

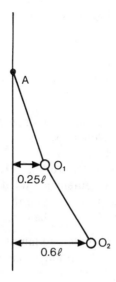

Fig. 9

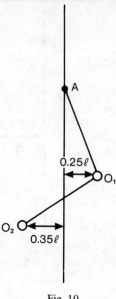

Fig. 10

The oranges should again exhibit SHM, but in anti-phase and with a shorter period. Again, the time T_2 for five oscillations should be recorded, and the required result should be that

$$\frac{T_1}{T_2} = \frac{n_2}{n_1} \approx 2.4$$

It may come as a surprise to see that a grossly simplified model can be validated so well by such crude apparatus. A further surprise may be achieved by eating one orange, and continuing with the next demonstration.

2.5 Free falling oranges
Apparatus: As in section 2.4, with an orange and, perferably, one rigid rod.

The object of this demonstration is to illustrate the concepts of energy conservation and Newton's first law by demonstrating aspects of the motion of a simple pendulum, and that of projectiles. The orange is suspended from A, as before, Fig. 11, the points B and C being vertically below A. The arrangement allows three modes of demonstration to be performed, each dependent upon the speed, u_o, of the orange at the point C. If l is the length BC, then the three modes are given by Green (1948),

(i) $u_o < \sqrt{2gl}$ two simple pendulums, or
(ii) $\sqrt{2gl} \leq u_o \leq \sqrt{5gl}$ motion in an arc of a circle, and then projectile motion, Fig. 12, or
(iii) $u_o \geq \sqrt{5gl}$ motion in a complete circle centred at B.

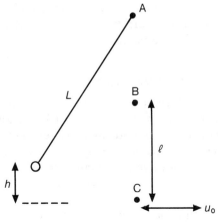

Fig. 11

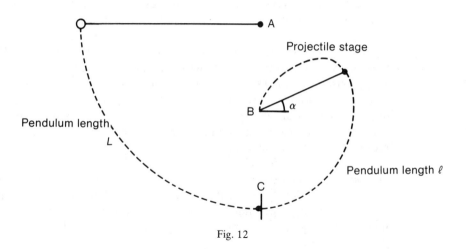

Fig. 12

The three modes can be illustrated in turn, the point B being set up as an obstruction using a horizontal rod. This is a similar approach to that described by Galileo in a very neat experiment (French, 1971). The arrangements for our apparatus, and a discussion of the experiment on this scale, were suggested by Trapp (1983).

Mode (i) is easily achieved by keeping the string taut and allowing the orange to fall, along a circular arc, from a height $h < l$, see Fig. 11. For mode (ii), the demonstration is most effective if the system is set up so that the orange passes through the point B during the projectile stage of the motion. It is straightforward to show that this may be achieved by releasing the orange from a position for which the string is horizontal and taut, and the point B fixed so that the ratio AC:BC is $(2 + \sqrt{3})/2$, see Fig. 12. Using a string of length $2m$, then BC $\approx 1.07m$ and the angle α is approximately $35.3°$. These values may sound rather precise, but it is not too difficult to set up the correct initial conditions and to obtain a successful conclusion. Finally, mode (iii) is carried out simply by giving the orange a sufficiently high speed, u_o at the point C.

3. FURTHER DEVELOPMENTS

The preceding discussion of practical demonstrations raises the question of which other areas of mechanics can be validated in the lecture room. Several standard undergraduate texts, such as Chester (1979), and Ramsey (1962) contain interesting examples, usually in the exercises which can be readily adapted for live presentation to suitable groups of students. Ideas gleaned from these sources include the following.

Orbits: motion under a central force, producing elliptic planar orbits, and also linking this with circular motion has exciting possibilities (Ramsey, 1962). It must be said in this case, where the interactive feature is not practicable, the TV/video presentation has some merit, for example, in discussions of Halley's Comet (The Open University, 1983).

Impact of spheres: the texts Synge and Griffith (1949), Ramsey (1962) describe experiments based on Newton's cradle which can either be used to verify conservation laws or, perhaps, to find the coefficient of restitution between two oranges. Practical considerations may suggest setting up an impact by means of two pendulums. Oranges or tennis balls can be used, and in the former case it is recommended that any fruit be placed in a deep freeze prior to the experiment!

Pulleys: arrangements of pulleys can be set up, not with the sophistication of Fletcher's trolley experiments, but more on the lines of the epic saga of the desperately unfortunate builder's labourer (Hoffnung, 1958). A suitable pulley rope and a household bucket can be used to good effect here.

Rockets: motion along a horizontal model railway track, using a small gas canister of the sparklets types containing a propellant, will provide an effective demonstration of Newton's third law.

Circular motion: the popular and original Open University TV presentation of hot wheels racing cars looping the loop is much more exciting in real time. However, in this case, the tracks used are not so popular as they were, and they may now be difficult to obtain.

The list is lengthy and additions simply require imagination and preparation on the part of the teacher. The most successful demonstrations usually evolve from discussions with colleagues. A chance remark, or communication out of context, is often the catalyst for the development of quite stimulating and worthwhile presentations. In trying these experiments, the lecturer will gain in confidence and experience and will find that ideas for developments and extensions will occur quite naturally. To quote a familiar proverb, nothing ventured, nothing gained.

The scope of such demonstrations can be extended by the use of photographic techniques to trace the path of a particle, but this often requires more sophisticated apparatus—a stroboscope and a polaroid camera. Projectile motion, elastic impact and normal modes are topics which have been covered in this manner in modelling courses at Brighton Polytechnic, and with the summer school practical element of the MST204 Mathematical Methods and Modelling course of the Open University. The guiding principles in the suggestions for interactive lecture demonstrations include the most important fact that they should be simple, effective and require no sophisticated equipment. There may be a temptation on the part of some teachers to emulate the physicist in this respect, but the authors believe that there is great merit in giving to the students demonstrations in which they themselves can participate and which they can repeat at home.

The demonstrations outlined in this chapter form an interesting and stimulating exercise, but they must not be used in isolation. They should be considered as an essential part of the whole modelling process, emphasising the need for model validation, and as an effective climax to the modelling which has taken place.

REFERENCES

Bajpai, A. C. *et al*. (1984). *Int. J. Math. Educ. Sci. Technol*., **15**, 781.
Bell, R. G. (1977). Private communication.
Chester, W. (1979). *Mechanics*. George Allen & Unwin.
Fendrich, R. (1982). MST204, Unit 24, Normal Modes, the Open University.
French, A. P. (1971). *Newtonian Mechanics*. Nelson.
Green, S. L. (1948). *Dynamics*. University Tutorial Press.
Harding, R. D. (1984). *Fourier Series and Transforms*. Adam Hilger.
Hoffnung, G. (1958). Speech to the Oxford University Union.
Lighthill, J. (Ed.) (1978). *Newer Uses of Mathematics*. Penguin.
Ramsey, A. S. (1962). *Dynamics*. Cambridge.
Synge, J. L., & Griffith, B. A. (1949). *Principles of Mechanics*. McGraw-Hill.
The Open University, Course MST 204 (1983) TV27, From Hastings to Halley—the Orbit of a Comet.
The Open University, Centre for Mathematics Education (1984) Visualising Mechanics.
Thorpe, M. A. (1984). Private communication.
Trapp, J. J. (1983). Private communication.

10

Continuous and Discrete Techniques in Mathematical Modelling

D. J. G. James

and

M. A. Wilson

Coventry (Lanchester) Polytechnic, UK

---

1. INTRODUCTION

With the growth in the availability of microcomputers for use in courses on mathematical modelling, it is likely that there will be an increased emphasis on the formulation of discrete models. This view may be supported from two different standpoints:

(i) Students, in general, find it more natural to use a discrete formulation when dealing with many situations; particularly situations where the associated data is only available in a discrete format.
(ii) Discrete models lead to easy programming for investigation on a computer and complexities incorporated within the model appear easy to introduce. Continuous models, however, become less tractable as more and more complexities are incorporated.

This chapter considers some of the problems students face when interpreting responses of discrete models which may not have associated parallels if the corresponding continuous model is adopted. It is tutorial in nature and provides suitable material for drawing out the issues within a classroom situation.

In order to highlight some aspects of its solution the standard limited resources growth model is first considered. This is included in order to provide relevant background material for the modelling exercise that follows.

2. LIMITED RESOURCES GROWTH MODEL

2.1 A typical situation
A typical situation is that depicted by a population growing within a closed environment which is subject to the availability of a limited supply of resources. The modelling of such a situation is well documented in the literature and will, therefore, not be investigated in detail here. It is characterised by a growth rate $\lambda(x, t)$ whose value at time t is dependent on the population size $x(t)$ at that instant. It is clearly not possible to specify the exact functional relationship between $\lambda(x, t)$ and $x(t)$ but it is realistic to assume that it has features as depicted graphically in Fig. 1.

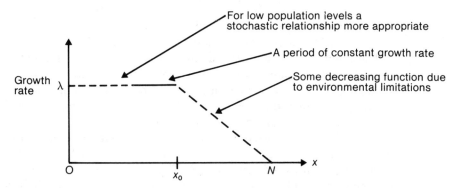

Fig. 1. Relationship between growth rate and population.

Taking the initial population level to be $x(0) = x_0$, where $x_0 < N$ and is sufficiently large to avoid the need for stochastic considerations, and assuming that the decreasing functional relationship between $\lambda(x, t)$ and $x(t)$ is linear then the situation may be modelled in a continuous form by the differential equation

$$\frac{dx}{dt} = \alpha x \left(1 - \frac{x}{N}\right), \tag{1}$$

or, in discrete form by the difference equation

$$x(n + 1) = x(n) + \alpha \left[1 - \frac{x(n)}{N}\right] x(n) \tag{2}$$

where α is a constant parameter.

2.2 The continuous model
The differential equation (1), subject to the initial condition $x(0) = x_0$ may be solved to obtain the explicit solution

$$x(t) = \frac{x_0 N}{x_0 + (N - x_0) e^{-\alpha t}} \tag{3}$$

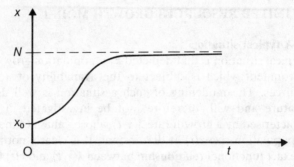

Fig. 2. Logistic growth curve.

which corresponds to the well known logistic growth curve of Fig. 2 and first proposed as a population model by Verhulst in 1938.

The point to note here is that the continuous model of equation (1), predicts that the population level $x(t)$ will approach the value N in an increasing monotonic manner independent of the values of both α and x_0. At this stage therefore, N may be interpreted as being the maximum population level sustainable within the environment.

As an interesting aside it could be pointed out to students that this information could have been obtained without actually obtaining the explicit solution (3) of equation (1). This is done graphically by drawing the phase-plane plot (see Fig. 3). By inspection of this plot it is clearly seen that for $0 < x_0 < N$ the population will always increase monotonically to the value N, which was the conclusion arrived at by considering the explicit solution.

2.3 The discrete model

Explicit solutions of nonlinear difference equations are not so readily obtained and it is unlikely that students would attempt to solve equation (2). Rather, the investigation of equation (2) will be undertaken

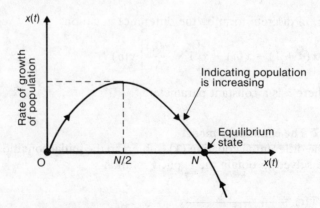

Fig. 3. Phase-plan plot for continuous model of equation (1).

BASIC program for LIMITED RESOURCE MODEL

```
10 lf$=chr$(13)
20 rem     -Limited resource model
30 rem      x(n+1)=x(n)+(a-bx(n))x(n)
40 dim x(100)
50 input "enter results-file name",n$
60 m=0
70 if n$<>"" then m=1
80 if m=0 then goto 100
90 open n$ as m
100 input "enter intercept",a
110 input "enter slope(magnitude)",b
120 input "enter initial population",x(0)
130 input "maximum time(integer)",mt
140 print #m,lf$;lf$;lf$
150 print #m,"        -Limited resource model";lf$;lf$
160 print #m,,"a =";int(1000*a+.5)/1000,,"b =";int(10000*b+.5)/10000
165 print #m,,"max. popn. =";int(a/b+.5)
170 print #m,,"x0 =";int(10*x(0)+.5)/10
180 print #m,,"tmax =";mt
190 print #m,,"time","population"
200 print #m,,0,int(10*x(0)+.5)/10
210 for j=1 to mt
220 x(j)=x(j-1)*(1+a-b*x(j-1))
230 print #m,,j,int(10*x(j)+.5)/10
240 next j
250 print #m,lf$;lf$
260 if m=0 then 280
270 close m
280 input a$
290 if a$="y" then 50
300 end
```

Fig. 4. BASIC program for investigating the discrete model (4).

using a small program on a microcomputer. The BASIC program of Fig. 4 outputs a sequence of values of $x(n)$ for the difference equation

$$x(n +) = x(n) + (a - b\,x(n))x(n) \tag{4}$$

where $a = \alpha$ and $N = a/b$ to match parameters with equation (2). When carrying out such a numerical investigation, students may come across responses having different characteristics. For instance, taking $x_0 = 1000$ and $b = 10^{-4}$ the responses of equation (4), corresponding to the parameter a having the three values 1, 2 and 3, are shown in Figs 5(i)–5(iii) respectively, where n has been mapped into appropriate t values. A more detailed discussion on such responses may be found in the text by Dorn and McCracken (1976).

2.4 Comment on the responses
Responses such as those of Figs 5(i) and 5(ii) are readily acceptable by students whilst a response such as that of Fig. 5(iii), which is rarely

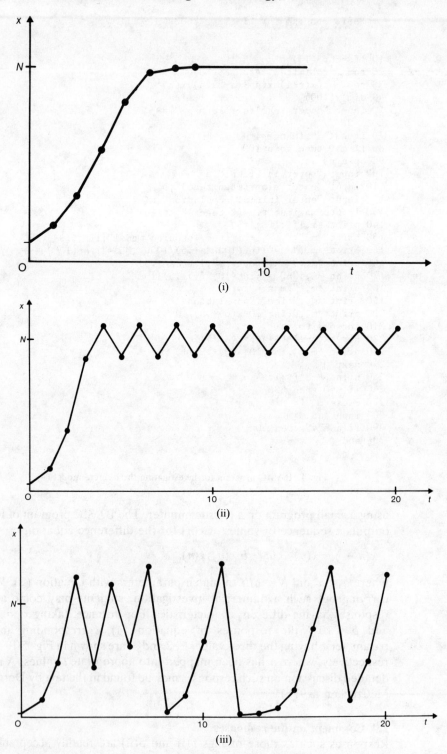

Fig. 5. Responses of discrete model.

obtained in practice, is most likely to lead to a search for errors in the program!

When students first meet such responses they do not relate them to differences in choice of parameter values and frequently argue with colleagues over differences. The fact that different values for the initial conditions can lead to different kinds of responses is not at all obvious to them.

It is easy to take the view that the various responses of section 2.3 are a result of poor numerical procedures for investigating the continuous model of equation (1). However, one must not assume that the continuous model is the most appropriate for replicating the real situation.

3. AN ILLUSTRATIVE SITUATION

3.1 Data set
In their text *A First Course in Mathematical Modelling* Giordano and Weir (1985) demonstrate the fitting of a logistic curve to Pearl's data (1927) on the growth of yeast in a culture. The method of fitting used requires that a reasonable estimate of the limiting population be available which may not be the case in practice. Pearl's data is reproduced here as columns (1) and (2) of Table 1, where column (1) denotes *time in hours* and column (2) denotes *the observed yeast biomass* and is used here as base data due to its quality of fit to the logistic curve.

3.2 Problem setting
The student is given the problem situation:
Here is some past population data taken at discrete points in time; forecast the future behaviour of the population.
The past data used is the yeast data of Table 1 or a subset of it. It is assumed that the problem situation is given to students having a reasonable knowledge of mathematics, including a knowledge of simple regression analysis. It is also anticipated that the students have available to them, as background aids, microcomputers and some useful mathematical packages.

3.3 Initial considerations
Population models may be dealt with in a variety of ways, but their basis, using the notation of section 2.1, is the growth model

$$\frac{dx}{dt} = \lambda(x,t)x$$

or an equivalent discrete form.

What should be stressed is that, in general, it is the form of $\lambda(x, t)$ as a function of t or, implicitly, as a function of x that is important; given a form for $\lambda(x, t)$ then much of the *guesswork* is removed thereafter. This important facet of modelling population situations is rarely stressed in

Table 1 Growth of yeast in a culture

(1) t	(2) Observed	(3) Increase	(4) Average	(5) Growth rate
0	9.6			
		8.7	13.95	0.623
1	18.3			
		10.7	23.65	0.452
2	29.0			
		18.2	13.10	0.478
3	47.2			
		23.9	59.15	0.404
4	71.1			
		48.0	95.10	0.505
5	119.1			
		55.5	146.85	0.378
6	174.6			
		82.7	215.95	0.383
7	257.3			
		93.4	304.00	0.307
8	350.7			
		90.3	395.85	0.261
9	441.0			
		72.3	477.15	0.152
10	513.3			
		46.4	536.50	0.086
11	559.7			
		35.1	577.40	0.061
12	594.8			
		34.6	612.10	0.057
13	629.4			
		11.4	635.10	0.018
14	640.8			
		10.3	645.95	0.016
15	651.1			
		4.8	653.50	0.007
16	655.9			
		3.7	657.75	0.006
17	659.6			
		2.2	660.70	0.003
18	661.8			

textbooks. Unfortunately, past data does not make it obvious as to what form should be taken for $\lambda(x,t)$ and it needs to be investigated in order to generate some ideas. In Table 1 the simple procedure of dividing *increase in population*, column (3), by *average population*, column (4), to get *growth rate* as a proportion, column (5), is demonstrated; there is obviously room for refinement on this simple approach. The results of plotting the value in column (5) against both *time*, column (1), and *population*, column (4), are then considered and a view taken as to a possible form for $\lambda(x,t)$. Not surprisingly, for this data, a plot of $\lambda(x,t)$

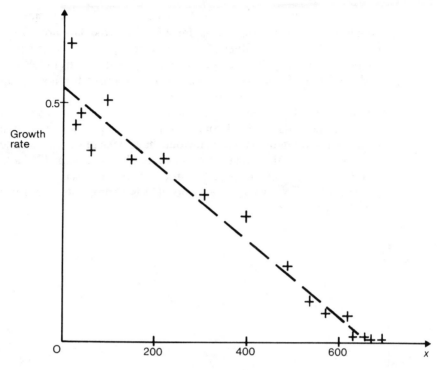

Fig. 6. Scatter diagram of growth rate against population and resulting regression line.

against x, shown in Fig. 6, may be accepted as linear giving $\lambda(x, t)$ in one of the two equivalent forms

$$\lambda = a - bx$$

or

$$\lambda = \alpha(1 - x/N)$$

Simple linear regression may be used to estimate the parameters of the line; the result of which is also shown in Fig. 6.

3.4 The model

This section is mainly concerned with the way a student tackles the problem of validation and then prediction. The more mathematical students may aspire along the continuous lines outlined in section 2.2 whilst the computer oriented students tend to opt for the discrete approach of section 2.3.

Using the data shown in Table 1 the least squares line indicated in Fig. 5 is

$$\lambda = 0.53129 - 0.0007951x \tag{5a}$$

or

$$\lambda = 0.53129 \, (1 - x/668) \tag{5b}$$

Taking the values of α and N from equation (5b) and substituting in equation (3) yields a response giving a fair fit to the data provided. On the other hand, using the values of a and b from equation (5a) in equation (4) yields a response which is a poorer, but still adequate, fit to the given data with the value of x_0 not being critical. Both responses are shown graphically in Fig. 7 and yield similar predictions.

The poorer fit of the discrete model response may be due to the way $\lambda(x, t)$ has been determined. Since, in the discrete model, increase operate from the last quoted population it would have been more appropriate in this case to determine the growth rate using the values in column (2), rather than those of column (4), of Table 1. Alternatively, the discrete model may be refined so that the $\lambda(x, t)$ of equation (5) is appropriate and a variety of ideas may be tried.

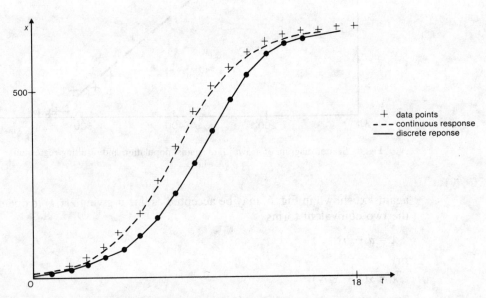

Fig. 7. Responses of continuous and discrete models for data of Table 1.

3.5 Limited data (sparse)

Should the given population data be rather less well defined and, say, only every third observation of population in column (2) of Table 1 be given then progressing along the lines outlined in sections 3.3 and 3.4 will result in changes in the observed resoponses.

Adopting the same procedure, as that applied in section 3.3 to the data of Table 1, to the reduced data yields a plot of $\lambda(x, t)$ versus x that strongly suggests linearity with the fitted line

$$\lambda = 1.410\,27 - 0.002\,103x \tag{6a}$$

or

$$\lambda = 1.410\,27\,(1 - x/671) \tag{6b}$$

having a correlation coefficient of −0.996 (which students will claim is nearly perfect!)

The responses of both the continuuous and discrete models of equations (3) and (4) are shown in Fig. 8, together with the test data. The 'omitted' data values are also indicated in the figure so that the effect of their omission may be judged. As may be observed the continuous model still shows an adequate fit and predicts a slow approach to the limiting population. The discrete model, however, shows a large discrepancy and produces a small oscillation attenuating to the limiting population similar in manner to that depicted in Fig. 5(ii).

If in the discrete case the determination of $\lambda(x, t)$ is refined along the lines discussed in the previous section then an even more extreme response than that depicted in Fig. 5(iii) is obtained. However, for this situation a plot of $\lambda(x,t)$ versus x does not indicate a linear relationship. Adopting a simplistic approach, such as making $\alpha = 2.5$ and keeping N at 671 yields a response which initially tracks the given data but soon becomes wildly oscillatory.

3.6 Limited data (truncated)

Here the situation when the given data is restricted to the early values is considered. The data set is assumed to be alternate values in columns (1) and (2) taken up to a time of 10 hours, thus yielding six observations of population values. Evaluating $\lambda(x,t)$, using the simple procedure outlined in section 3.3, yields five values for $\lambda(x,t)$ which look linear when plotted against x. (In this case a linear assumption is also not unreasonable when $\lambda(x,t)$ is plotted against t and an investigation of this situation is left as an

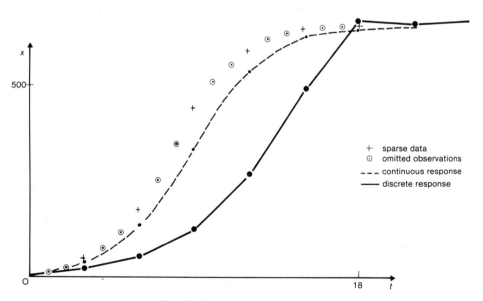

Fig. 8. Responses for sparse data.

exercise for the reader.) Taking the view that the second value is an aberration the linear fit

$$\lambda = 1.037 - 0.00150x$$

is obtained. The corresponding response of the continuous model, equation (3), provides a moderate fit, whilst that of the discrete model, equation (4), provides a poor fit to the observed data. If the values of $\lambda(x,t)$ are determined, along the lines suggested in section 3.4, using the last quoted population value then the plots $\lambda(x,t)$ suggest that it is a linear function of time but not of x; suggesting therefore that this is not a limited resources situation. The 'casual' fit of

$$\lambda(x,t) = 2.06 - 0.385t$$

when used in the discrete model leads to a response which is a good fit to the given truncated adata (even better than the continuous model response) but given our knowledge of what eventually happens its predictions are surprising. The responses are illustrated in Fig. 9.

4. CONCLUSIONS

The chapter considers some of the issues a teacher may have to face up to when advising students using a discrete approach to a mathematical modelling exercise. The main points to which attention must be drawn are

(i) One must not restrict the student's approach as there is no clear indication in advance as to whether a continuous or discrete approach is the most appropriate for a particular situation.

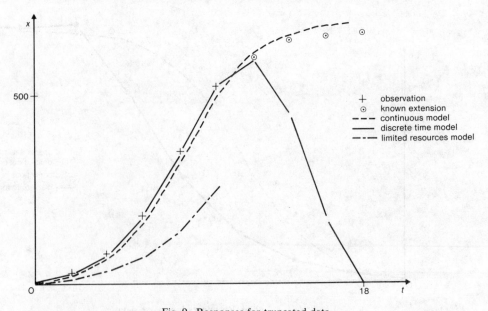

Fig. 9. Responses for truncated data.

(ii) A discrete model should not be looked upon as being a crude numerical approximation to a continuous model.
(iii) During the investigatory stage discrete models require two aspects to be considered:
  (a) effects arising directly from issues relating to the problem situation;
  (b) effects attributable to the properties of difference equations; for example small departures in parameter values can lead to responses with distinctly different characteristics.
  This contrasts with the continuous model where only the consequences of (a) have to be considered.
(iv) In a discrete model formulation small enhancements are not so effectively introduced as they may, for example, require a complete re-estimation of parameters. Consequently, considerably more thought should be given to the initial design of discrete models than is often the case.

REFERENCES

Dorn, W. S. & McCracken, D. D. (1976). *Introductory Finite Mathematics with Computing*. J. Wiley.

Giordano, F. R. & Weir, M. D. (1985). *A First Course in Mathematical Modelling*. Brooks/Cole Pub. Company, California.

James, D. J. G. & Steele, N. C. (19   ). Harvesting a renewable resource, J. Math. Modelling Teachers, **2**, 3, 34–42.

Pearl, R. (1927). The growth of population, *Quart. Rev. Biol.*, **2**, 532–548.

Verhulst, P. F. (1938). Notice sur la loi que la population suit dans son accroissement, *Correspondance Mathematique et Physique*, **10**, 113–121.

11

Mechanics via Cosmology

P. T. Landsberg
University of Southampton, UK

SUMMARY

The most successful teachers manage to engage the enthusiasm of their students, be it by clarity, or by relevance to an important contemporary issue, or by showing that students can now obtain interesting results by themselves. A case will be made out for importing much needed enthusiasm into the teaching of classical mechanics by the use of ideas from cosmology. This can be made to exhibit all three of the above characteristics. In this chapter some examples will be given on how this somewhat unusual application of mechanics can be handled. The relation between energy equations and equations of motion can be exhibited, using Hubble's law. One can also deduce the simpler cosmological models in this way. Critical density, deceleration parameter, a singularity theorem, and some black hole properties, can be explained quantitatively within Newtonian mechanics. Provided some caveats are introduced, and one is not too ambitious, results are found that are in agreement with, but do not depend on, general relativity.

The chapter contains one introductory section and six mathematical appendices which illustrate our topic.

1. INTRODUCTORY SECTION

Mathematical modelling means that one takes a given problem provided by our surroundings, and replaces it by a more clearly defined situation which is mathematically tractable. The approximations are then contained in the transition "reality → model". This procedure has been followed in classical mechanics for centuries. Just as human subjects become grey over the years, it seems the very age of the subject of mechanics has enveloped its aura at school also with a certain greyness. The extensible string, the rigid disc, the smooth surface—they do not make you think of vibrant life, of exciting discovery; I am afraid they make you think of the teacher in his study, pencil in hand. There is nothing wrong with this; I am such a person myself. I like my study. However, I want to project applied mathematics to be something so exciting that the study itself becomes a seat of excitement.

There must be several ways of doing so. One which I have advocated for some time is to use cosmology as a vehicle (Landsberg, 1963). This seems on first reflection a hard thing to do, since the general relativity theory, on which modern cosmology is based, will remain beyond the school syllabus for decades. However, there were published in 1934 some papers which lifted the curtain by showing that, over a certain large area, Newtonian cosmology gives results which are essentially the same as those of relativistic cosmology (McCrea & Milne, 1934). Of course one has to exercise care. The expansion of space itself is not part of the Newtonian picture, since in the latter a system of galaxies expands into a *pre-existing* space. Space curvature is also unknown in Newtonian theory. Remarkably enough, however, the main cosmological models of the universe—which fortunately are the simple ones—arise in Newtonian theory almost exactly as they do in relativistic theory. Thus a book about the Newtonian approach—following in the steps of E. A. Milne and Bill McCrea—seemed desirable and was written (Landsberg & Evans, 1977). It is within the reach of a good sixth former, but one should be able to do better, by adopting *consistently* the point of view of a sixth former. I am engaged in this task at this time.

For the present purpose it must suffice to give examples. Below I give six considerations from Newtonian cosmology which lead to conclusions which are identical with those one would derive from the general theory of relativity. We shall use classical mechanics, and this suggests that it could be taught via the medium of cosmology.

In the first example the Hubble parameter $H(t)$ is introduced. It can be backed up by some historical remarks and anecdotes, and it can be used to teach kinematics (the study of the motion of bodies without considering the causes of these motions). Applications, properly conceived, lead one directly to cosmological models, which can be studied and generalised beyond what I do here. Newton's laws of motion and of gravitation come next. Combined with Hubble's law, one finds the main differential equation of cosmology: equation (12) or (14). As an application, one arrives at the escape velocity and an estimate of the so-called critical density of cosmological theory. It is but a small step now to model the black hole and gravitational collapse.

It is somewhat novel to approach this subject in a strictly Newtonian frame of mind. However, there are a good many books on cosmology at all levels of sophistication. We cite a few which give some mathematics without being too advanced for the present context (Berry, 1976; Rowan-Robinson 1981; Narlikar 1983; Layzer 1984).

APPENDIX 1. KINEMATICS: THE LAW OF EDWIN HUBBLE

1. Events $(\mathbf{r},t)$ are noted by the reception of a light pulse by an observer at $(\mathbf{r}_0,t_0)$. The times of occurence are always corrected for the time it

takes light to reach the observer:

$$t_0 = t + |\mathbf{r}_0 - \mathbf{r}|/c \tag{1}$$

This holds for all models in classical and special relativity mechanics.
2. For all galaxies i at all times t the position from some origin is

$$\{\dot{\mathbf{r}}_i(t) = H(t)\mathbf{r}_i(t)\} \to \{\mathbf{r}_i(t) = \mathbf{r}_i(0)\bar{R}(t)\}, \frac{\bar{R}(t)}{\bar{R}(t_0)} = \exp\int_{t_0}^{t} H\, dt \tag{2}$$

3. The deceleration parameter is

$$q_i \equiv -\frac{\ddot{r}_i r_i}{\dot{r}_i^2} = -\frac{\dot{H} r_i + H^2 r_i r_i}{H^2 r_i^2} = -\frac{\dot{H}}{H^2} - 1 \quad \text{(indep. of } i\text{)} \tag{3}$$

Problem (a) (Kinematics). Prove that galaxies do not overtake each other.
Solution (a) Use (2). If at t^* $r_A(t^*) = r_B(t^*)$, then

$$\{\bar{R}(t^*)r_A(0) = \bar{R}(t^*)r_B(0)\} \to \{r_A(0) = r_B(0)\} \to \{r_A(t) = r_B(t)\}$$

at *all* times. So galaxies are the same distance from the origin at one time only if they are so at *all* times.
Problem (b) Find q for a steady-state universe.
Solution (b) Use (3). In a steady-state H is the same at all times, whence $q = -1$. Then by (2)

$$\bar{R}(t) = \bar{R}(0)\exp(Ht) \tag{4}$$

Problem (c) (Simple Differential Equation). Find the dependence on time of $\bar{R}(t)$ and $H(t)$ if q is time-independent ($\neq -1$).
Solution (c) If q is a constant and A and B constants of integration,

$$\left\{\frac{\ddot{\bar{R}}}{\bar{R}} = -q\frac{\dot{\bar{R}}}{\bar{R}}\right\} \to \left\{\bar{R}^q \dot{\bar{R}} = \frac{A}{q+1}\right\} \to \{\bar{R}^{q+1} = At + B\}. \tag{5}$$

Thus

$$\{(q+1)\bar{R}^q\dot{\bar{R}} = A\} \to \left\{H = \frac{1}{(q+1)(t + B/A)}\right\}. \tag{6}$$

Note if $R = 0$ at $t = 0$, then $B = 0$.

APPENDIX 2. APPLICATIONS OF KINEMATICS: COSMOLOGICAL MODELS

The problems (b) and (c) (q = const.) have the following applications.

(i) *The steady-state model $q = -1$.*
This model is due to H. Bondi, T. Gold and F. Hoyle (1948) and can be derived from the perfect cosmological principle that neither history nor geography matters. Thus the average density of matter and the Hubble constant are time-independent.

(ii) *The Milne model* $q = 0$. $R(t) = At + B$, $H(t) = (t + B/A)^{-1}$ (7)
If galaxies have constant velocities then if $r_i(0) = 0$

$\mathbf{r}_i(t) = \mathbf{v}t$

So by (2)

$$\bar{R}(t) = \frac{r_i(t)}{r_i(t_0)} = \frac{t}{t_0}, \quad H(t) = \frac{\dot{r}_i(t)}{r_i(t)} = \frac{v}{vt} = \frac{1}{t};$$ (8)

At $t = 0$ there is a big-bang singularity. The Hubble time H^{-1} is the age of the universe.

Ever-expanding models tend to go over into the Milne model, since the matter is thinned out to make for zero gravitational deceleration.

(iii) *Einstein–de Sitter model (1932)* $q = \frac{1}{2}$. Use problem (c) with $B = 0$.

$$\bar{R}(t)^3 = A^2 t^2, \quad H(t) = 2/3t$$ (9)

The age, t, of the universe is two-thirds times the Hubble time H^{-1}. Some astronomical observations support $q = \frac{1}{2}$. Several models approach this one as $R \to 0$.

APPENDIX 3. RELATION BETWEEN AN ENERGY EQUATION AND AN EQUATION OF MOTION

The energy equation

$$T = \frac{1}{2} \sum_i m_i \dot{r}_i(t)^2 = \frac{1}{2} \left[\sum_i m_i r_i(0)^2 \right] \dot{\bar{R}}(t)^2 = A \dot{\bar{R}}(t)^2$$ (10)

$$V = -G \sum_{i<j} \frac{m_i m_j}{r_{ij}(t)} = -G \left[\sum_{i<j} \frac{m_i m_j}{r_{ij}(0)} \right] \frac{1}{\bar{R}(t)} \equiv -\frac{B}{\bar{R}(t)}$$ (11)

$$-\frac{E}{A} = \frac{B}{A\bar{R}} - \dot{\bar{R}}^2 = \frac{B}{A\rho\bar{R}}[\rho - \rho_c]$$ (12)

$$\rho_c = \frac{A}{B} \rho \bar{R} \dot{\bar{R}}^2$$ (13)

For $\rho > \rho_c$, $E < 0$ and the model collapses.
For $\rho < \rho_c$, $E > 0$ and the model expands for ever.

The equation of motion

$$-\frac{B}{A\bar{R}^2}\dot{\bar{R}} - 2\dot{\bar{R}}\ddot{\bar{R}} = 0 \quad \text{or} \quad \ddot{\bar{R}} = -\frac{B}{2A\bar{R}^2}$$ (14)

i.e.

$$q = -\frac{\ddot{\bar{R}}\bar{R}}{\dot{\bar{R}}^2} = \frac{B}{2A\bar{R}\dot{\bar{R}}^2} = \frac{\rho}{2\rho_c}$$ (15)

Thus for $q \leq \frac{1}{2}$ ($\rho \leq \rho_c$) the models expand forever. This includes the Einstein–de Sitter model, which has $\rho = \rho_c$. For $q > \frac{1}{2}$ ($\rho > \rho_c$) there is a recontraction.

The $\ddot{R}$ equation shows that $\bar{R}$ is *concave* towards the t-axis. Thus there exists a time t_s such that $R(t_s) = 0$ and therefore $\rho(t_s)$ diverges. This is a singularity theorem.

Starting with the equation of motion (3.5), the energy equation (3.3) can be obtained by integration.

APPENDIX 4. APPLICATION OF THE ENERGY EQUATION: ESCAPE VELOCITY AND THE CRITICAL DENSITY

The total energy of a body of mass m is in Newtonian terms its kinetic energy plus its potential (gravitational) energy due to the presence of a heavy gravitating particle of mass M at distance r

$$E = -\frac{1}{2}mv(r)^2 - \frac{GMm}{r} = \frac{1}{2}mv(\infty)^2 \tag{16}$$

Here the velocity is measured in a frame of reference in which the heavy body is at rest, and G is Newton's gravitational constant. It has been assumed that the particle can escape to infinity with velocity $v(\infty)$. If this is only just possible, then $v(\infty) = 0$. In any case the condition for escape from a spherical surface of radius r is

$$v(r) \geq \sqrt{(2GM/r)} \equiv v_e(r) (\approx 33.7 \text{ Mach for the earth}) \tag{17}$$

This is the famous escape velocity.

If M is due to a uniform distribution of matter of density ρ and of radius in excess of r, then only the mass within the radius r is effective in providing a gravitational pull. Furthermore, if the cloud of matter is expanding according to Hubble's law, the inequality is

$$H^2 r^2 \geq \frac{2G}{r} \frac{4\pi}{3} \rho r^3 \tag{18}$$

whence $\rho \leq 3H^2/8\pi G \equiv \rho_c (\approx 5 \times 10^{-30} \text{ gm cm}^{-3}$ if $H^{-1} = 6 \times 10^{17}$ s)*

This is the condition for an expanding model universe in Newtonian (as well as in general relativistic) theory. For the Einstein–de Sitter model $\rho = \rho_c$.

APPENDIX 5. APPLICATION OF THE EQUATION OF MOTION. COLLAPSE TO A BLACK HOLE

For any particle of mass M there exists a distance r_o from it such that for $r < r_o$ no escape is possible because $v_e(r_o)$ in (17) is the velocity c of light *in vacuo*. Working in c.g.s. units, with $M_\odot$ denoting the mass of the sun.

$$r_0 = \frac{2GM}{c^2} = \frac{2 \times 6.7 \times 10^{-8}}{9 \times 10^{20}} \frac{2 \times 10^{33}}{M_\odot} \sim 3 \frac{M}{M_\odot} \text{ km}$$

For a uniform distribution of matter of density ρ the above holds if M is interpreted as $(4\pi/3)\rho\, r_0^3$. In intermediate cases when the object is confined to a region of radius $r < r_0$ it is called a black hole and not even light can escape from it.

If a test particle resides on the surface of a spherically symmetric object collapsing from zero velocity at a radius r at time $t = 0$, then its equation of motion is

$$\ddot{x} = -GMx^{-2}, \text{ i.e. } \dot{x}^2 = 2GM(x^{-1} - r^{-1}) \approx 2GMx^{-1} = r_0 c^2 x^{-1}$$

if $x \ll r$. Hence $|\dot{x}| = c$ when the black hole radius $x = r_0$ is reached. Another integration gives the time t_0 needed for collapse to this radius

$$\frac{2}{3}(r^{3/2} - r_0^{3/2}) \sim c r_0^{1/2} t_0.$$

Hence

$$t_0 \sim \frac{2}{3}\frac{r}{c}\sqrt{\frac{r}{r_0}} \sim \frac{2 \times 7 \times 10^5 (\text{Km})}{3 \times 3 \times 10^5 (\text{Kms}^{-1})}\sqrt{\frac{70 \times 10^4}{3}} \sim 750 \text{ s}$$

The numerical work applies to the collapse of an object which is the size of the sun.

APPENDIX 6. NON-CONSERVATION OF MASS: CREATION RATE OF MATTER IN THE STEADY-STATE MODEL

If $v(t) = 4\pi r^3/3$ is the volume of space bounded by specified galaxies at time t, the volume bounded by them at time $t + \Delta t$ is

$$v(t + \Delta t) = v(t) + 4\pi r^2 h \equiv v(t) + \Delta v$$

where h is the thickness of the new shell of matter created by expansion. If H is the present value of the Hubble parameter $\approx (6 \times 10^{17} \text{s})^{-1}$ then

$$\Delta v = 4\pi r^2 \times Hr\, \Delta t .$$

Since Hubble's law does not allow the overtaking of one galaxy by another, no galaxies enter the volume from the outside. The generation rate per unit volume, G, of matter can therefore be obtained from the present matter density $\delta_0 \approx 10^{-26}$ Kg m^{-3} which must fill the newly created volume Δv. Hence

$$G = \frac{\delta_0 \Delta v}{v \Delta t} = \frac{\delta_0 \times 4\pi r^3 H \Delta t}{(4\pi/3) r \Delta t} = 3\delta_0 H.$$

The mass of a hydrogen atom is $\approx 2 \times 10^{-27}$ Kg, a year is 3.2×10^7 sec so that

$$G = \frac{3 \times 10^{-26}}{6 \times 10^{17}} = 5 \times 10^{-44} \text{ Kg m}^{-3}\text{s}^{-1}$$

$$= \frac{5 \times 10^{-44} \times 3.2 \times 10^{10}}{2 \times 10^{-27}} = 8 \times 10^{-7} \text{ H atoms m}^{-3} \text{ per 1000 years.}$$

This is roughly one hydrogen atom in a cube of side 100 m in 1000 years.

REFERENCES

Berry, M. (1976). *Principles of Cosmology and Gravitation*. Cambridge University Press.
Landsberg, P. T. (1963). *Math. Gaz.*, **47**, 101.
Landsberg, P. T. & Evans, D. A. (1977). *Mathematical Cosmology: An Introduction*. Oxford University Press.
Layzer, D. (1984). *Constructing the Universe*. W. H. Freeman.
McCrea, W. H. & Milne, E. A. (1934). *Q. Jl. Math.*, **5**, 73.
Milne, E. A. (1934). *Q. Jl. Math.*, **5**, 64.
Narlikar, J. V. (1983). *Introduction to Cosmology*, Jones & Bartlett.
Rowan-Robinson, M. (1981). *Cosmology*. Clarendon Press, Oxford.

12

Discrete Linear System Modelling Techniques

K. V. Lever
University of East Anglia, UK

SUMMARY

Information technology makes increasing use of discrete techniques, such as digital computers, for information representation and processing. As a result, greater emphasis is falling on the use of discrete models in the design process. This shows up in changes in the mathematics curriculum of Information System Engineers as a shift towards discrete mathematics with less emphasis placed on continuous mathematics as such, and more placed on the relationships between continuous and discrete.

1. INTRODUCTION: INFORMATION SYSTEMS AT UEA

This chapter is concerned with the mathematics component in the formation of Electronic Systems Engineers, and the way in which inexorable trends towards discrete systems in information technology need to be matched by a corresponding emphasis on discrete mathematics.

The views advocated by the author are the result of participation in a three year redevelopment of the Electronic Systems Engineering programme at the University of East Anglia. This redevelopment has been part of a wider rationalisation of information technology interests in the University, which has resulted in the foundation of a new School of Information Systems incorporating undergraduate, postgraduate, research and consultancy activities in accountancy, business information systems, computer science, computer systems engineering, as well as electronic systems engineering.

This reorganisation has been accompanied by a radical reappraisal of virtually all the subject matter in each of the five undergraduate programmes offered in the School. Each of the new curricula emerging from this process is a consensus of requirements generated in a variety of ways: educational or vocational prerequisites for individual courses, cross-sectoral commonality to ensure optimal managment of resources, and the desire to preserve the interdisciplinary traditions of the University. In this chapter, attention is drawn to the process of formulating requirements for a mathematics curriculum to support the formation of Electronic Systems Engineers.

Electronic systems engineering is envisaged (Fig. 1) as the combination of three cognate disciplines: computing, communication and control. These share common roots in the use of electronic techniques, with a preference for microelectronic implementation. This in turn embodies a comprehension and exploitation of physical principles—mainly (but not exclusively) the physics of the solid state. But there is another unifying framework, that of mathematical modelling. Formal modelling techniques play an increasingly important role in the design of modern information systems at the professional level (see, for instance, Cattermole & O'Reilly, 1984), and this necessitates a similar emphasis in undergradulate curricula. It is not always appreciated in schools and sixth-form colleges, for example, just how much modern system engineering relies upon abstract mathematics for modelling purposes. Accordingly, system engineering is much more a natural career for a mathematical specialist than, say, a physicist. The discussion that follows is aimed at teachers of mathematics in the later years of secondary and the early years of tertiary education. The objective is a modest one: to demonstrate that the concepts of discrete mathematics, as propounded for instance in Lipschutz (1976), form a natural foundation for one topic of major importance in the design of information systems, namely linear system theory. In fact, the applicability of discrete mathematics goes far beyond the single example chosen, and penetrates deep into the foundations of the subject. An account of these foundations can be found, for instance, in Shannon's seminal exposition of the Mathematical Theory of Communication (Shannon & Weaver, 1964):

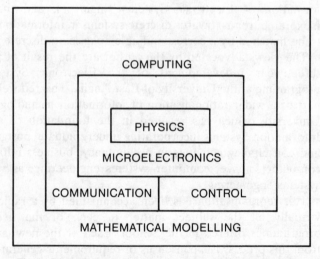

Fig. 1. Disciplinary structure of electronic systems engineering.

most of that material should be accessible to a good sixth-former or first-year university student.

2. SIGNALS AND SYSTEMS

A system may be regarded as an operator $\mathscr{S}$ processesing an input signal x to produce an output signal y. For definiteness, we may think of x and y as voltages, modelled as functions of time as in Fig. 2.

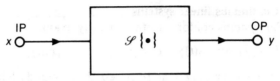

Fig. 2. System diagram.

There are four signal categories to consider: the time variable (denoted by t) and the voltage value (denoted by v) can be either continuous or discrete (sampled) as in Fig. 3.

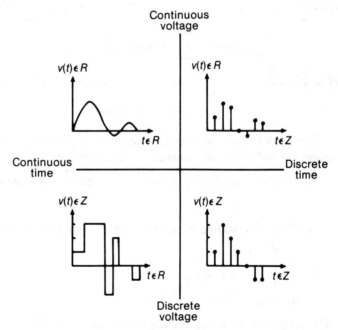

Fig. 3. Signal categories.

The system input–output relation is depicted in various ways, for example:

$$x \xrightarrow{\mathscr{S}} y \qquad (1)$$

$$\mathscr{S} : x \longrightarrow y \qquad (2)$$

$$\mathscr{S}\{x(t)\} = y(t) \qquad (3)$$

Signals x and y are considered as elements of the same signal (vector) space V, and the system operator as an element of a space of operators $\mathcal{O}$. For simplicity we restrict our attention to the case when both signals and systems are deterministic, and we concentrate on the subspace of linear systems.

3. LINEAR SYSTEMS

3.1 Continuous linear systems

Linear systems conform to the linearity axiom

$$\mathcal{L}\{c_1 x_1(t) + c_2 x_2(t)\} = c_1 \mathcal{L}\{x_1(t)\} + c_2 \mathcal{L}\{x_2(t)\} \tag{4}$$

where $\mathcal{L} \in \mathcal{O}$; $x_1, x_2 \in V$ and $c_1, c_2 \in F$ the field underlying V. Usually, but not always, $F = \mathbb{R}$ or $\mathbb{C}$. In the physical sciences this property is known as the principle of superposition, and in system theory it is usually depicted pictorially as in Fig. 4.

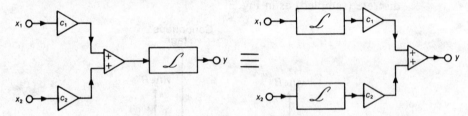

Fig. 4. The linearity axiom.

The modelling of linear systems starts by appealing to familiar notions of primitive linear operations such as the following.

multiplicative scaling by a constant (amplification/attenuation)

$$y(t) = c.x(t) \tag{5}$$

multiplicative scaling by another time-varying function (modulation)

$$y(t) = c(t).x(t) \tag{6}$$

time-translation (delay)

$$y(t) = x(t - T) \tag{7}$$

The diagrammatical equivalents are shown in Fig. 5.

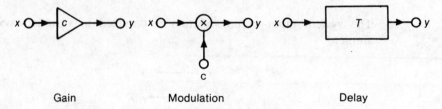

Gain   Modulation   Delay

Fig. 5. Primitive linear operations.

Note that the addition of a constant (or another signal) is a nonlinear operation. But the addition of a scaled delayed version of the original signal is linear:

$$y(t) = x(t) + h.x(t - T) \tag{8}$$

This can be extended to cover the case of a finite linear combination of delayed inputs,

$$y(t) = \sum_i h_i x(t - T_i) \tag{9}$$

and then, by taking the limit, to the case of an infinite linear combination of input signals delayed by infinitesimal increments

$$y(t) = \int h(t') x(t - t') \, dt \tag{10}$$

The most general case can be obtained by making the change of variable $t' \to t - t'$ and allowing the constant coefficients h to vary with time t.

$$y(t) = \int h(t,t') x(t') \, dt' \tag{11}$$

Thus the operator may be considered as a generalised linear transform. Writing it as a linear functional

$$\mathcal{H}\{\cdot\} = \int h(t,t') \cdot (t') \, dt' \tag{12}$$

emphasises the way the transform operates on arbitrary input signals (denoted by the vacancy symbol '·'). The function $h(\cdot,\cdot)$ is called the system kernel (not to be confused with kernel used in the sense of null-space). The general linear operator is time-varying. This is obvious from the system diagram model, Fig. 6.

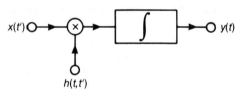

Fig. 6. Time-varying linear system.

The time-variation is also manifest by diagnosing the system behaviour by means of an input signal chosen for its simplicity—the Dirac delta-function impulse $\delta(t' - T)$ (a generalised function), located at time $t' = T$. For this case the output, the impulse response, is

$$y(t) = \int h(t,t') \, \delta(t' - T) \, dt' = h(t,T) \tag{13}$$

In general, the output $y(t)$ will be a different function for different values of T, and the system is therefore time-varying. That is, the system is characterised by a family of impulse responses, parametrised by T.

The special case for which the behaviour of $\mathcal{H}$ does not fluctuate with T is of particular importance: time-invariant linear systems are called filters and are characterised by a translational impulse response

$$h(t,t') = h(t - t') \tag{14}$$

and the corresponding system operator is called convolution,

$$y(t) = \int h(t - t') x(t') dt' \quad (15)$$

and is given the special symbol *

$$y = h * x \quad (16)$$

In addition to convolutional time-domain modelling, time-invariant linear systems admit to further exploration via system invariants or eigensignals. These are signals that suffer only a (complex) change of amplitude, that is scaling, in passage through the system. They provide a method of diagnosing system behaviour complementary to impulse-testing. The eigensignals of a convolutional operator are complex exponential functions of time, parametrised by an additional variable f.

$$e_f(t) = \exp(j2\pi ft) \quad (17)$$

The corresponding eigen value is λ_f, or in functional notation $H(f)$. That is

$$\mathcal{H}\{\exp(j2\pi ft)\} = H(f) \cdot \exp(j2\pi ft) \quad (18)$$

$H(f)$ is given by the equation

$$H(f) = \int \exp(-j2\pi ft) h(t) \, dt = \mathcal{F}\{h(t)\} \quad (19)$$

were $\mathcal{F}$ denotes the Fourier transform. Since the eigensignal set is linearly independent and complete (within some suitably defined signal space) the Fourier transform (and its inverse $\mathcal{F}^{-1}$) may be regarded as orthogonal transformations, or rather unitary transformations in view of the complexity of the underlying field.

The parameter f may be regarded as a frequency, and $H(f)$ is accordingly termed the frequency response. Note that this approach shows that the use of the Fourier transform is not a matter of convenient choice. Given that the system is linear and time-invariant, the eigensignals can only be the complex exponentials, and the eigenvalues are inexorably determined as a Fourier transform.

The multiplicative effect of the system on the eigensignals generalises (because the system is linear) to all signals representable as a linear superposition of eigensignals. That is, if

$$x(t) = \int X(f) \exp(j2\pi ft) df \quad (20)$$

and

$$y(t) = \int Y(f) \exp(j2\pi ft) df \quad (21)$$

then

$$y(f) = H(f) \cdot X(f) \quad (22)$$

The functions $X(f)$ and $Y(f)$ represent the frequency content of the signals $x(t)$ and $y(t)$, respectively, at each individual frequency f. They are, of course, the spectra of their corresponding signals, so that $\int \cdot \exp(j2\pi ft) df$ is the inverse Fourier transform $\mathscr{F}^{-1}$, and we have arrived at the convolution theorem:

$$
\begin{array}{ccccc}
x & \longrightarrow & x*y & \longleftarrow & y \\
\mathscr{F} \updownarrow \mathscr{F}^{-1} & & \mathscr{F} \updownarrow \mathscr{F}^{-1} & & \mathscr{F} \updownarrow \mathscr{F}^{-1} \\
X & \longrightarrow & X.Y & \longleftarrow & Y
\end{array}
\qquad (23)
$$

Thus the Fourier transform canonicalises the convolution operator, simplifying its relatively complicated multiply—and—integrate character into a purely multiplicative form.

3.2 Discrete linear systems

In this case the time variable is discretised, that is sampled, so that the signal can now be regarded as a sequence or vector indexed by an integer $n \in Z$.

$$x(nT) = x(n); \qquad y(nT) = y(n) \qquad (24)$$

$$\{x(n)\} \leftrightarrow \mathbf{x}; \qquad \{y(n)\} \leftrightarrow \mathbf{y} \qquad (25)$$

The generalised linear operator is also discretised

$$\mathscr{H}[x(m)] = y(n) = \sum_{m \in Z} h(n, m) x(m) \qquad (26)$$

which is easily recognised as a matrix transformation

$$\mathbf{y} = \mathbf{H}\mathbf{x} \qquad (27)$$

with

$$[\mathbf{H}]_{nm} = h(n, m) \qquad (28)$$

provided the signal sequences are regarded as column vectors.

Specialisation to the time-invariant case models a discrete filter by means of a convolution matrix whose rows and columns are translated versions of the unique impulse response

$$[\mathbf{H}]_{nm} = h(n,m) = h(n-m) \qquad (29)$$

for example,

$$
\mathbf{H} = \begin{bmatrix}
 & & \cdot & \cdot & \cdot & \\
 & & \cdot & h(-1) & h(0) & h(1) & \cdot \\
 & \cdot & h(-1) & h(0) & h(1) & \cdot \\
\cdot & h(-1) & h(0) & h(1) & \cdot & \\
 & \cdot & \cdot & \cdot & &
\end{bmatrix}
\qquad (30)
$$

Discrete convolution can be modelled as above by matrix operations, or algebraically as follows.

$$y(n) = \sum_{m \varepsilon Z} h(n - m) \times (m) \stackrel{\Delta}{=} h * x(n) \qquad (31)$$

This is recognisable as a generalisation of the scalar product between two vectors. In this case there is a time-reversal and time shift of n samples, as shown in Fig. 7.

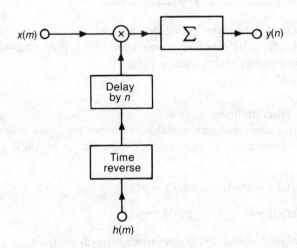

Fig. 7. Convolution processor.

Simple as it seems, this sum-of-products computation is extremely important, as it underlies much innovative engineering in signal processing for example, correlation detection and pattern recognition.

As microprocessor speeds improve it becomes increasingly more feasible to implement useful processors of this type by means of real-time software systems, so that the appropriate model is an algorithm. For instance, the flow-chart for the above filter is given in Fig. 8, and a notional program in Fig. 9.

A graphical model is a useful aid to comprehension: the usual approach is to use transparent overlays to illustrate the reverse, shift, multiply and add structure of convolution. Computer animation of the process is also helpful. This is complemented by an exploded tabular version of the algorithm using numerical values, as for example in Fig. 10.

This approach usefully accentuates the triviality of the convolution operation—it is merely long multiplication without the complication of the carry.

The architectural model corresponding to the tableau of Fig. 10 is known as transversal or feedforward or finite-impulse-response (FIR) filter. This is shown in Fig. 11.

There is also a matrix-algebraic model corresponding to signal synthesis and system characterisation by system invariants. In the discrete case we

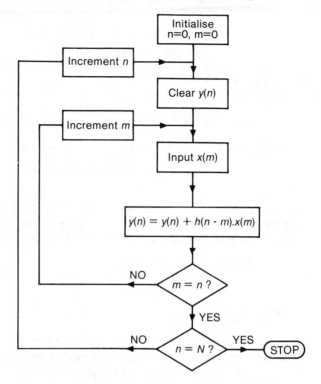

Fig. 8. Convolution algorithm flowchart.

are considering eigenvectors of the **H**-matrix: these are the sampled exponentials

$$\mathbf{e}_p = [\exp(j2\pi p 0/N), \exp(j2\pi p 1/N), \ldots, \exp(j2\pi pq/N), \ldots]^T \quad (32)$$

where $\mathbf{H}\mathbf{e}_p = \lambda_p \mathbf{e}_p$ expresses the eigenvector–eigenvalue properties. (33)
Here N is the length of the input and output vectors, that is

$$0 < p, q < N - 1 \text{ and } \dim V = N$$

.
.
.

    for $n = 0$ to N
        $y(n) = 0.0$
        for $m = 0$ to n
            $y(n) = y(n) + h(n) - m) * x(m)$
        next m
    next n

.
.
.

Fig. 9. Convolution algorithm program

$$h * x = y$$
$$[2, 4, 6]] * [1, 0, -1] = [2, 4, 4, -4, -6]$$

	$h(0) = 2$	$h(1) = 4$	$h(2) = 6$		
$x(0) = 1$	2.1	4.1	6.1		
$x(1) = 0$		2.0	4.0	6.0	
$x(2) = -1$			2.(−1)	4.(−1)	6.(−1)
	2	4	4	−4	−6

Fig. 10. Tabular convolution.

so that **H** is an $N \times N$ matrix. Apart from a $1/N$ scaling factor, the eigenvalues are obtained by transformation of the impulse response vector **h** by the adjoint (transpose conjugate) of the matrix of eigenvectors

$$\lambda_p = [\mathbb{E}^+ \mathbf{h}]_p \tag{34}$$

where

$$[\mathbb{E}]_{pq} = [\mathbf{e}_p]_q \tag{35}$$

The unitary (complex) matrices $\mathbb{E}^+$ and $\mathbb{E}$ are clearly the discrete counterparts of the forward and inverse Fourier transforms, respectively. Using the customary functional notation we have

$$\lambda_p = H(p) = \sum_q \exp(-j2\pi pq/N) h(q) \tag{36}$$

—which is the discrete frequency response. There is, of course, a discrete version of the convolution theorem.

One of the features to emerge from this approach is the fact that the matrix entries consist of the N primitive roots of unity

$$E_{pq} = \exp(j2\pi pq/N) \tag{37}$$

This emphasises the periodicity of the model. The usual cartwheel representation in the complex plane (Fig. 12) highlights another property: that there is an underlying group (and subgroup) structure.

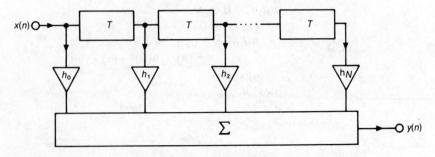

Fig. 11. Convolution architecture.

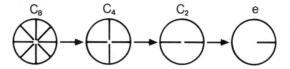

Fig. 12. Cyclic group structure.

The fact that $C_2\alpha$, the cyclic group of order 2^α ($=N$) decomposes into subgroups implies the existence of fast algorithms for computing the discrete Fourier transform. The subgroup decomposition induces a decomposition of the $\mathbb{E}$ matrix into a product of $\alpha = \log_2 N$ submatrices

$$\mathbb{E} = \mathbb{E}_1 \mathbb{E}_2 \ldots \mathbb{E}_\alpha \tag{38}$$

This appears to involve a greater computational workload than the N^2 operations required to compute $\mathbf{X} = \mathbb{E}^+\mathbf{x}$ (or $\mathbf{x} = \mathbb{E}\mathbf{X}$) directly. In fact each $\mathbb{E}_j$ is considerably sparse, and the total workload is actually reduced to about $N\alpha = N \log_2 N$ operations. This is a saving of $N : \log_2 N$—which is enormous for even modest values of N (e.g. $\simeq 100:1$ for $N = 1024$). Thus the celebrated Fast Fourier Transform (Rabiner & Gold, 1975), which is usually regarded as a spectacular example of innovative algorithm design in computing science, or hardware/software engineering in signal processing, is primarily the exploitation of innate mathematical structure. Similar ideas permeate related areas of computer science: reduction of workloads from N^2 to $N \log_2 N$ by structured algorithm design is common in data-processing tasks such as searching, sorting, shuffling and merging (Page & Wilson, 1978).

The recognition of group structure provides other insights into linear system behaviour. For example the matrices $\mathbb{E}$ and $\mathbb{E}^+$ may be regarded as modal matrices (Ayers, 1974), which diagonalise the convolution matrix

$$\mathbb{E}\mathbf{H}\mathbb{E}^+ = \mathrm{diag}\,\mathbf{H}$$

where $\mathbf{H}$ is the frequency response vector. Linear systems can be cascaded as in Fig. 13.

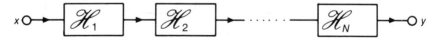

Fig. 13. Cascaded linear systems.

This corresponds to forming the product of the corresponding matrices as follows.

$$\mathbf{H} = \mathbf{H}_N \ldots \mathbf{H}_2 \mathbf{H}_1 \tag{39}$$

However, each $\mathbf{H}_i$ can be written as

$$\mathbf{H}_i = \mathbb{E}^+ \mathrm{diag}\,\mathrm{H}_i \mathbb{E} \tag{40}$$

so that

$$\mathbb{H} = \mathbb{E}^+ \operatorname{diag} H \tag{41}$$

where, by virtue of the cancellation of $n - 1$ $\mathbb{E}\,\mathbb{E}^+$ terms we have

$$\operatorname{diag} H = \prod_{i=1}^{N} \operatorname{diag} H_i \tag{42}$$

Since diagonal matrices form an abelian (semi)group, so too do linear systems. Furthermore, the non-existence of inverses which prevents the establishment of the full group property, can be related, via determinantal analysis, to the singularity of the diagonal matrix and this admits to a system-theoretic interpretation (zeros in the frequency response) having important practical ramifications.

The point about analysis (and synthesis) of discrete linear systems is that it can proceed by way of matrix algebra, which is usually perceived as being more comprehensible than the integral transform algebra corresponding to the continuous case—partly because appeal can be made to familiar geometrical analogies, but also because the discrete case is more amenable to illustration with numerical examples and computer aids.

4. A DESIGN EXAMPLE: DATA SCRAMBLERS AND DESCRAMBLERS

Data scramblers (Bylanski & Ingram, 1980), (not to be confused with coders for data compression, error control or encryption) are used in digital communication systems to ensure that the transmitted signals contain sufficient timing information to permit the establishment and maintenance of synchronism in the receiver without recourse to providing a side-channel wasteful of valuable transmission capacity. The system scenario is depicted in Fig. 14.

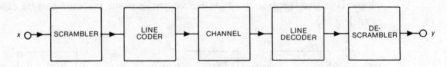

Fig. 14. Scrambled data transmission.

The scrambler usually operates on the binary source data prior to line coding, and its purpose is to break up long runs of 1's or 0's which, resembling dc, would lack strong components at the pulse repetition frequency needed for synchronisation. The device consists of shift-registers providing delay and exclusive—OR gates providing modulo-2 addition. An example is given in Fig. 15.

The question is, given such a scrambler, how does one design the inverse processor (descrambler) located in the receiver? The quickest way is to recognise that this is a transversal feedforward filter in which both signal

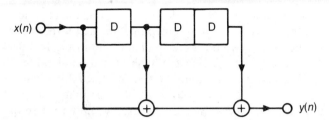

Fig. 15. Scrambler architecture.

values and convolutional weights are restricted to GF(2). Apart from this feature, system analysis follows the usual procedure. The time-domain model is

$$y(n) = h \bigcirc x(n) \quad \text{where } h = [1, 1, 0, 1] \tag{43}$$

i.e.

$$y(n) = x(n) \oplus x(n-1) \oplus x(n-3) \tag{44}$$

The spectral representation obtained by taking Z-transforms is:

$$Y(z) = X(z) \oplus z^{-1} X(z) \oplus z^{-3} X(z) \tag{45}$$

so that the system response is

$$H(z) = \oplus z^{-1} \oplus z^{-3} \tag{46}$$

The descrambler must have system response $G(z)$ inverse to $H(z)$

$$G(z) = H^{-1}(z) = [1 \oplus z^{-1} \oplus z^{-3}]^{-1} \tag{47}$$

which can be interpreted in terms of its input–output representations as

$$Y(z) [1 \oplus z^{-1} \oplus z^{-3}] = X(z) \tag{48}$$

so that

$$Y(z) = X(z) \oplus z^{-1} Y(z) \oplus z^{-3} Y(z) \tag{49}$$

Here we have made use of the identity of addition and subtraction in GF(2). The time-domain model is

$$y(n) = x(n) \oplus y(n-1) \oplus y(n-3) \tag{50}$$

This defines the recursive, infinite impulse response (IIR) architecture of the descrambler shown in Fig. 16.

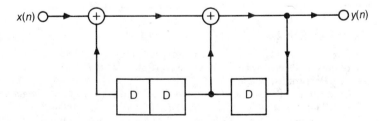

Fig. 16. Descrambler architecture.

Other issues can be decided by exploiting the commutativity of operators and their inverses (without invoking linearity). In practice errors picked up on the channel could be recirculated within the feedback structure of the descrambler. So it is sensible to site the transversal feedforward processor in the receiver rather than the transmitter. Although some error extension may occur, it is unlikely to be as large as that of a feedback descrambler. Because the two devices are inverse, we may reverse their roles—with no effect in the noiseless case, and with probable benefit in practice.

5. CONCLUSIONS: CHANGING THE EMPHASIS FROM CONTINUOUS TO DISCRETE

Although the foregoing example was selected from the author's own field of interest (signal processing), the overall picture is thought to be similar in a number of disciplines contributing to the design of information systems. This swing towards the discrete representation of information is obviously driven by the use of discrete systems (such as digital computers) to implement information processing. Accordingly, discrete mathematics is a more natural approach, for information system engineers and information scientists, than the traditional one based on continuous mathematics. This is not to say that continuous mathematics is now redundant. On the contrary, it is essential for the design of continuous systems such as analogue filters and classical control and communication systems. And there are several cases where a discrete signal or system might have a continuous representation (e.g. the Z-transform), and a satisfactory comprehension of system behaviour is only possible if the associated continuous models can be competently handled. Specifically, there is a need for continuous mathematics covering the theory of limits, convergence, differential equations, integration and complex variable. However, these topics should be taught in such a way as to place strong emphasis on the relationship between the continuous and the discrete: perhaps the only satisfactory way to do this is to introduce the theory of generalised functions at an early stage, albeit with an engineering emphasis, as in Zemanian (1965). This presents real pedagogical challenges, which can only be met by adopting a mature axiomatic approach in which mathematics is considered not as a collection of supporting techniques, but as the theoretical *foundation* of information technology.

REFERENCES

Ayers, F. (1974). *Matrices*, Schaum Outline Series. McGraw-Hill, p. 163.
Bylanski, P. & Ingram, D. G. W. (1980). *Digital Transmission Systems*. Peter Peregrinus, p. 268.
Cattermole, K. W. & O'Reilly, J. J. (Ed.) (1984). *Mathematical Topics in Telecommunications. Volume 1: Optimisation Methods in Electronics and Communications. Volume 2: Problems of Randomness in Communication Engineering*. Pentech Press.

Lipshutz, S. (1976). *Discrete Mathematics*. Schaum Outline Series, McGraw-Hill.
Page, E. S. & Wilson, L. B. (1978). *Information Representation and Manipulation in a Computer*. Cambridge University Press.
Rabiner, L. R. & Gold, B. (1975). *Theory and Application of Digital Signal Processing*. Prentice-Hall, p. 356.
Shannon, C. E. & Weaver, W. (1964). *The Mathematical Theory of Communication*, University of Urbana Press. also: July and October 1948. *Bell System Technical Journal*.
Zemanian, A. H. (1965). *Distribution Theory and Transform Analysis*. McGraw-Hill.

13

Improving Driver Comfort in Motor Vehicles

A. Norcliffe, G. G. Rodgers and M. M. Tomlinson
Sheffield City Polytechnic, UK

SUMMARY

The BSc Applied Science degree course at Sheffield City Polytechnic has recently been expanded by the inclusion of a subject area entitled Computing and Applicable Mathematics. This subject area was introduced in response to the need for scientists who are trained not only in the traditional analytical and experimental skills but also in computer disciplines.

We have found case studies invaluable in the teaching of this subject area, and in this chapter we report in detail on one such case study which we use with our second year students.

In Section 1 we describe the nature of the Computing and Applicable Mathematics course and we set the case study in its teaching context.

In Section 2 we describe the background, theory and organisation of the case study, and in Section 3 we report on the lessons we have learned in running the case study.

1. INTRODUCTION

The BSc Applied Science degree at Sheffield City Polytechnic is one of three degrees which comprise the Degree Programme in Science, the others being BSc Applied Chemistry and BSc Metallurgy and Microstructural Engineering. All these are thin sandwich courses which include two six-month periods of industrial experience within the four-year course duration.

Part I of the courses consists of the first year, spent in the Polytechnic, followed by the first period of industrial experience. In Part I the three courses are taught totally in common. Each student chooses a study pattern made up of units in the subject areas of biology, chemistry, computing and applicable mathematics, physics, metallurgy/materials science. The units chosen depend upon the student's educational background, interests and chosen degree course. This common first year allows considerable flexibility in that the students' progression on a degree course/subject combination is finally decided at the end of Part I. After Part I the three degree courses continue independently.

Part II consists of a further year in the Polytechnic, followed by the second period of industrial experience. Students on BSc Applied Science study two subjects from the five identified above. These subjects are developed further in Part III which consists of the final year in the Polytechnic.

Computing and Applicable Mathematics, which we abbreviate CAM, with apologies for any confusion with computer-aided manufacture, serves a dual role in the Degree Programme in Science. First, in Part I, it provides the foundation of mathematics, statistics and computing for all students irrespective of their degree course/subject combination. Secondly, it stands as a subject in its own right within the framework of the BSc Applied Science degree.

2. THE CAM COURSE

There are two CAM course units in Part I

CAM 1.1, Basic mathematics, statistics and computing
CAM 1.2, Industrial applications of mathematics and computing

CAM 1.1 is compulsory for all students, while CAM 1.2 is an optional unit designed to illustrate examples of the applications which a student may meet on an industrial experience placement.

The CAM 1.1 unit provides core material in mathematics, statistics, probability and computing, including programming in a high level language. In teaching the course emphasis is placed on the applications of the material in science. The mathematical background of the students on the course is varied. All have O-level mathematics or its equivalent. Some have A-level mathematics. A considerable number enter the course with the intention of studying CAM as a major component of their course. Others are extremely apprehensive about mathematics in any form. For this reason the CAM 1.1 course is taught with sympathy for the needs and abilities of the students by, for example, forming teaching groups according to the students' chosen major science subjects. Towards the end of this unit case studies are introduced to illustrate the nature of mathematical modelling and to integrate material covered in the course.

The case studies in the CAM 1.1 course form the major part of the component of the course entitled Scientific Systems Modelling. The students are introduced to the nature, role and scope of modelling in science and the case studies allow them to follow the modelling process through the stages of model formulation in the representation of a scientific system, model development and implementation with associated computer programming, model testing and modification, and finally model application in the generation of appropriate output. Subject areas used in the case studies include chemical kinetics applied to consecutive first order chemical reactions, and the oscillatory behaviour of a damped mechanical system such as a vehicle suspension.

In Part II of BSc Applied Science course a student may study CAM and one other science subject. A very popular subject combination is CAM and Physics, but generally all four of the alternative subjects are combined with CAM. The course develops core material in mathematics, statistics, numerical analysis and computer methods, but considerable emphasis is placed on Scientific Systems Modelling. The methodological basis for the modelling of scientific systems is developed and a framework is provided for the integration of the core course material through application example. The computer methods component concentrates on data capture, storage and manipulation, these being essential areas in the development of computer-based models in science. As part of the Scientific Systems Modelling component the students are required to undertake two major case studies. One case study in this area is based on the development of an appropriate data structure for a computer program which determines the winter heating requirements of a building. The other case study is in the area of mathematical modelling with associated computer program development, and is the case study discussed in detail in this chapter. The Part II case studies differ from those in Part I in that in Part I the students are carefully led through a fully written up case study, stage by stage. In Part II the emphasis is placed on the students' own efforts, through both group and individual working.

In Part III additional core material is introduced in the areas of mathematics, numerical analysis and computing, but the major emphasis is placed on Scientific Systems Modelling. The subject matter includes control theory and optimisation. Again case studies play a major role in the course and these are based on the modelling and control of more complex scientific systems, and the interfacing of scientific and computer systems. A major part of the students' Part III assessment is a Project. The Project is selected from those offered by the contributing departments. Many projects span departments and reflect collaborative departmental research activities. We feel that the work carried out in the case studies in CAM is useful in preparing the students for the final year project.

Clearly we regard case studies as an essential component of the CAM teaching. In the following sections of this chapter we report in detail on our experience in running a case study in Part II of the course.

3. THE CASE STUDY

3.1 Background to the case study

The Applied Physics department at Sheffield Polytechnic is involved in a comprehensive research program into the mechanical properties of polymer foams and their relationship to foam structure. From the point of view of the Applied Science Degree, there are several problems arising from this research which have the potential for illustrating real world applications of physics, mathematics and computing as part of case study and final year project material. In choosing material for a Part II CAM case study, the criteria we wished to meet were that the material involved an interesting real world application, was of an appropriate standard of difficulty for second year students, and was sufficiently self-contained to make a worthwhile study within the fairly constrained time available towards the end of the Part II course. On the basis of these criteria, the problem chosen was that of looking at the mathematical modelling of vehicle seating—a study which had originally formed part of a collaborative venture between the Polytechnic, B and K Laboratories UK, and JCB Research Ltd, and which is outlined in the next section.

Vehicle seat modelling
In the last twenty years, polyurethane (PUR) foam has replaced rubber latex as a material for vehicle seat cushions. For manufacturing and cost reasons, seats composed of a system of springs and cushions have gradually been replaced by full depth foam sets. With the trend towards enhanced suspension stiffness and lighter vehicles, the properties of foam seats in isolating the passenger/driver from transmitted vibration have become of great importance in vehicle design. The mechanical properties of a PUR foam may be altered at the manufacturing stage (Benson *et al.*, 1973; Patten *et al.*, 1974; Pollart *et al.*, 1974), and hence knowledge of the appropriate mechanical properties for a particular vehicle-seat combination would enable optimisation of the seat for increased passenger/driver comfort. Clearly a useful tool in this optimisation process would be a mathematical model capable of predicting the way in which a seat manufactured from a foam with known mechanical properties will transmit vibrations. Several workers have performed experimental measurements of vibration transmission characteristics, both in the field and in the laboratory (Griffin, (1978); McNulty & Douglas, 1982; Collier *et al.*, 1983). If a mathematical model can be developed which will successfully reproduce experimental data then this is a good first stage in developing a model for predictive purposes. It is this first stage that the case study is concerned with.

From the several models that have been put forward to reproduce the response of the human body to cyclical oscillations (Band *et al.*, 1970; Payne & Wright-Patterson, 1979; International Standard, 1981), a model that has been found to reproduce experimental results for the transmission of seat vibrations reasonably well is one based on the work of Band *et al.* (1970). This model, shown schematically in Fig. 1, is composed of 3 masses and 3 resilient elements with complex moduli and is the model concentrated on in this study.

The labels attached to the parts of the model are intended only as a very rough guide, and for schematic simplicity the viscera is shown above the head and shoulders rather than in its physical position.

A resilient element with complex modulus $E^*(\omega)$ has dynamic stress $\sigma(\omega)$ and dynamic strain $\varepsilon(\omega)$ related by the expression

$$\sigma(\omega) = E^*(\omega)\varepsilon(\omega)$$

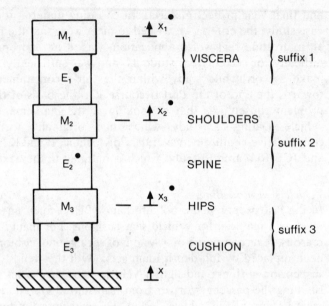

Fig. 1. Schematic diagram of Person-seat model.

where ω = angular frequency of vibration of the element. $E^*(\omega)$ thus plays a similar role to an impedance in complex AC circuit theory, allowing for a phase shift between ε and σ. Investigations by Snowdon (1968), have led to the conclusion resilient elements are more suitable for modelling body parts and cushions (which behave as distributed systems) than spring-dashpot arrangements (i.e. lumped systems), which have been widely used in this field. The stiffness $S^*(\omega)$ of a resilient element is given by

$$S^*(\omega) = kE^*(\omega)$$

where k is a constant depending on system geometry. As a first approximation the frequency dependence of S^* can be neglected, and we can express S^* in the form

$$S^* = S(1+j\delta)$$

where δ, known as the loss tangent, is the tangent of the phase shift between the displacement of a resilient element and the force applied to it.

If the underside of the foam cushion undergoes a displacement X^*, as a result of road vibration, and the three masses have displacements X_1^*, X_2^* and X_3^* as shown in Fig. 1, then with the suffices indicated in Fig. 1, the forces F_1^*, F_2^* and F_3^* on the resilient elements are given by

$$F_1^* = S_1^*(X_2^* - X_1^*)$$
$$F_2^* = S_2^*(X_3^* - X_2^*)$$
$$F_3^* = S_3^*(X^* - X_3^*)$$

The measure taken of the transmission of vibrations is the TRANSMISSIBILITY T, defined as

$$T = \left| \frac{X_3}{X} \right|,$$

that is the acceleration felt by the passenger/driver compared with the acceleration at the lower end of the seat. By calculating T from the model, the variation of T with ω can be investigated for varying values of M_i, S_i and δ_I ($i = 1,2,3$) and compared with experimental plots of T versus ω.

3.2 The case study programme

With the information outlined above, the specific objectives of the case study were that the student should

(i) derive an expression for T in terms of the M_i, δ_i and S_i,
(ii) write a program to produce plots of T versus ω for varying values of the M_i δ_i and S_i,
(iii) compare theoretical curves for T produced by the program with experimental plots,
(iv) consider the role of modelling in the design of vehicle seating, and discuss the advantages and limitations of the particular model used in this study.

The case study was run during a four-week period, consisting of a 2-hour session each week, and, in the final week or so, additional supervised time on a BBC ECONET system each evening, for the students to use as they needed. At the beginning, each student was issued with a booklet containing in essence (although in a rather more expanded form) the information presented in the last section, together with a timetable for the operation of the case study and detail of assessment requirements. To gain worthwhile experience within the time available, it was felt necessary to place a fairly close structure on the work to be done at each stage. What follows is a detailed breakdown of the progression of the study over the four weeks.

Week 1
The first hour was taken up with the presentation of the vehicle seat problem to the students, and an outline of the objectives of the case study (we are much indebted to Cole Brothers department store in Sheffield for the gift of an old tailors' dummy—christened Bertha by us—which, after suitable modification, greatly facilitated the explanation of the mathematical model of the person–seat system). The students were then asked to divide themselves into groups of three or four and, by way of getting started, have a look at the forced SHM equation to familiarise themselves with the idea of finding non-transient solutions using complex phasors. With the idea of complex numbers fresh in their minds, the groups

were given the task of deriving the equations of motion for the person–seat model in the form

$$M_1 X_1^* = S_1^*(X_2^* - X_1^*)$$
$$M_2 X_2^* = S_2^*(X_3^* - X_2^*) - S_1^*(X_2^* - X_1^*)$$
$$M_3 X_3^* = S_3^*(X^* - X_3^*) - S_2^*(X_3^* - X_2^*)$$

and, from these, in the time before the next session, deriving an expression for the transmissibility T in terms of the M_i, δ_i and S_i. The method suggested for finding the tranmissibility was as outlined below.

Since X^* is the vibrational input to the seat, if this is of angular frequency ω then (disregarding transient responses) X_1^*, X_2^* and X_3^* will be of the form

$$X_1^* = X_1 e^{j(\omega t + \varphi_1)}$$
$$X_2^* = X_2 e^{j(\omega t + \varphi_2)}$$
$$X_3^* = X_3 e^{j(\omega t + \varphi_3)}$$

where φ_1, φ_2 and φ_3 are constant phase shifts. The acceleration and displacements are thus related as follows, $\ddot{X}_1^* = -\omega^2 X_1^*$, $\ddot{X}_2^* = -\omega^2 X_2^*$ and $\ddot{X}_3^* = -\omega^2 X_3^*$. The equations of motion may now be written in the matrix form

$$\mathbf{AX} = \mathbf{z}$$

where

$$\mathbf{X} = \begin{pmatrix} X_1^* \\ X_2^* \\ X_3^* \end{pmatrix}, \quad \mathbf{z} = \begin{pmatrix} 0 \\ 0 \\ S_3^* X^* \end{pmatrix}$$

and

$$\mathbf{A} = \begin{pmatrix} S_1^* - M_1\omega^2 & -S_1^* & 0 \\ -S_1^* & S_1^* + S_2^* - M_2\omega^2 & -S_2^* \\ 0 & -S_2^* & S_3^* + S_2^* - M_3\omega^2 \end{pmatrix}$$

Then, solving for X_3^* by Gaussian elimination (or otherwise) leads to

$$T = \frac{S_3^*\{(S_1^* - M_1\omega^2)(S_1^* + S_2^* - M_2\omega^2) - (S_1^*)^2\}}{(S_1^* - M_1\omega^2)\{(S_1^* + S_2^* - M_2\omega^2)(S_1^* + S_3^* - M_3\omega^2) - (S_2^*)^2\} - S_1^{*2}(S_2^* + S_3^* - M_3\omega^2)}$$

This expression can now be put in a more convenient form for computation by substituting $S_1^* = S_1(1 + j\delta_1)$, $S_2^* = S_2(1 + j\delta_2)$, $S_3^* = S_3(1 + j\delta_3)$ and separating out the real and imaginary parts of the numerator and denominator. This process, although in principle straightforward, is somewhat tortuous, and leads to the expression

$$T = \left(\frac{R_N^2 + I_N^2}{R_D^2 + I_D^2} \right)^{1/2}$$

where

$$R_N = M_1M_2S_3\omega^4 - [M_1S_2S_3(1 - \delta_2\delta_3) + S_1S_3(1 - \delta_1\delta_3)(M_1 + M_2)]\omega^2$$
$$+ S_1S_2S_3(1 - \delta_1\delta_2 - \delta_1\delta_3 - \delta_2\delta_3)$$
$$I_N = M_1M_2S_3\delta_3\omega^4 - [M_1S_2S_3(\delta_2 + \delta_3) + S_1S_3(\delta_1 + \delta_3)(M_1 + M_2)]\omega^2$$
$$+ S_1S_2S_3(\delta_1 + \delta_2 + \delta_3 - \delta_1\delta_2\delta_3)$$
$$R_D = -M_1M_2M_3\omega^6 + [S_2(M_1M_2 + M_1M_3) + S_1(M_1M_3 + M_3M_3)]\omega^4$$
$$- [(M_1 + M_2 + M_3)S_1S_2(1 - \delta_1\delta_2)]\omega^2 + R_N$$
$$I_D = [S_2\delta_2(M_1M_2 + M_1M_3) + s_1\delta_1(M_1M_3 + M_2M_3)]\omega^4$$
$$- [(M_1 + M_2 + M_3)S_1S_2(\delta_1 + \delta_2)]\omega^2 + I_N$$

Week 2
At the beginning of the second session, the correct expression for T (although not the derivation) were made available to the students to enable each group to check the expressions they had derived.

The purpose of the second session was then to develop a BBC BASIC program to input values of the M_i, S_i and δ_i, calculate T for a range of frequencies between 1 and 20 Hz, and plot a graph of T versus ω using the Acornsoft Graphs and Charts package. The second session was conducted using the same student groups as the first session, and the programs written at this stage were first attempts to be run, debugged and further developed on an individual basis later.

For the remainder of the case study the students would be working individually and so, to round off the group working phase, each group was required to submit a report by the third week. The specification given to the students for the reports was as follows.

Group reports should contain an introduction to the case study and include the following.

(1) The role of person–seat models.
(2) The assumptions behind the particular model considered.
(3) The derivation of the equations of motion in the complex–number formulation used.
(4) The solution of these equations giving the transmissibility, T, explicitly as a function of ω and the parameters specifying the system.
(5) A preliminary version of a program/algorithm for evaluating T for specified input values of ω and the system parameters.

Each member of a group was given the same mark, and this counted as 40% of the total assessment for the case study.

Weeks 3 and 4
After producing a working version of the preliminary program developed in week 2, the primary task for each student for the remainder of the case study was to compare graphs of T versus ω from the model with experimental data. Experimental plots using accelerometer measurements have been obtained by Griffin (1978) and McNulty *et al.*

(1982) for person–seat systems in several types of vehicle. Four experimental plots were used in the case study, these being the plots shown in Fig. 2 for a small low-cost 5-door car, a small van, a 12-seater light bus and a JCB earth-moving vehicle.

One plot was issued to each student, ensuring that those who had worked together in the same group received different plots. Each student

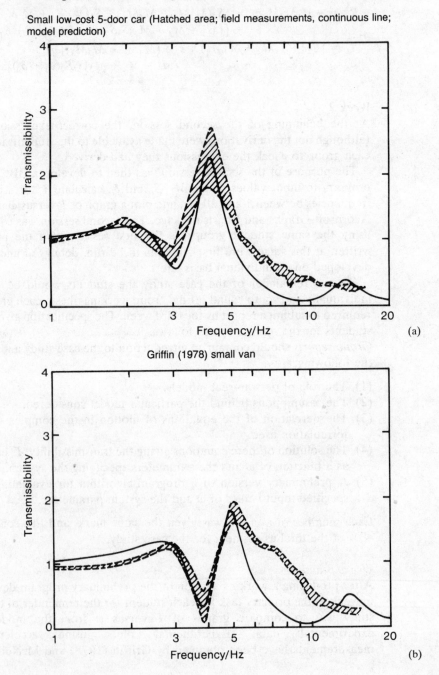

Improving Driver Comfort in Motor Vehicles

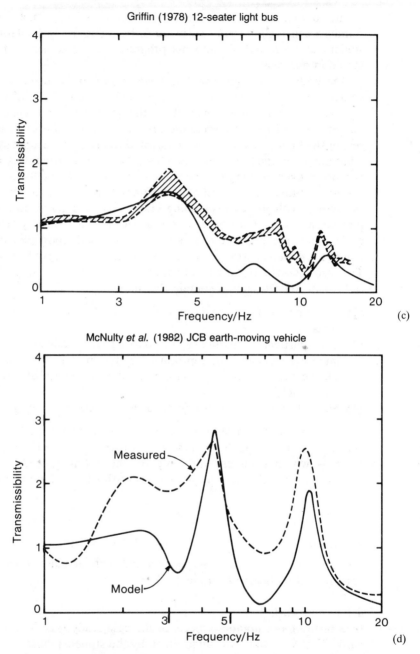

Fig. 2. Experimental plots for the transmissibility in four types of vehicle.

then experimented with values of the M_i, δ_i and S_i (within given ranges) to produce a plot from the model which matched as closely as possible with their experimental plot. The method of optimising the fit between experimental and model plots was left to the individual, although it was suggested that it would be helpful if both plots were displayed on the screen simultaneously.

The computing was carried out on an Econet system which enabled the standard charts and graphs software to be made easily available to each student and allowed all students programs and graphical output to be stored in one disk.

One hour of the final two hours was reserved for feedback with the students and staff involved in the case study. The purpose of this was to discuss what had been achieved by the modelling and what were the limitations, and also to discuss the operation of the case study itself. The rest of the final two sessions were devoted to using the Econet system and additionally, in the last week and a half, the Econet system was staffed for two hours each evening for the students to use as they felt necessary.

Once a student found a set of parameters that produced a satisfactory fit, the screen with the corresponding graph was dumped on to a file to be printed out later on a printer not attached to the Econet system. Finally each student was required to produce an individual report shortly after the end of the 4-week period to present the results of the curve fitting exercise. The instructions given for this final report were as follows.

Individual Reports should contain the following.

(1) A listing of a working program for producing frequency response plots.
(2) The particular frequency response plot obtained by measurement/experiment to be reproduced by the model.
(3) The best fit frequency response plot obtained (by trial and error) using the model.
(4) Suggestion for *systematically* producing a best fit to observed plots.

This report was worth 60% of the final assessment for the case study. It was expected that, in the course of instruction (3) and (4) above, the students would comment not only on the fitting procedure but also consider the possibilities and limitations of this type of modelling.

4. EXPERIENCE OF RUNNING THE CASE STUDY

We now report on how the students and staff actually coped with the person–seat modelling exercise.

4.1 Group work

One of the more notable aspects of the case study was the use of group work. The case study was undertaken by the students during a four-week period with only two hours per week class contact (out of a total of 18 hours per week) and with very demanding deadlines to be met. The only way, it seemed to us, that the students were going to complete the work required, on time, was by working in groups, in the first two weeks. The idea was that by pooling their resources the students would be able to complete the initial modelling phase and be able to produce the outline program that was required for week 3.

The 21 students formed themselves into three groups of 4 and three groups of 3. The students were asked to choose their own groups and told us afterwards, in the one hour feedback session, that they had preferred having the freedom to do this. It was pointed out that often in the real world they would have no say into which team they were assigned.

All groups met their first deadline, producing their group reports on time. The six group marks were as follows:

Group 1	(4 students)	34/40	85%
Group 2	(3 students)	29/40	73%
Group 3	(4 students)	25/40	63%
Group 4	(4 students)	22/40	55%
Group 5	(3 students)	23/40	58%
Group 6	(3 students)	25/40	63%
	Average:	26.3/40	65.8%

The marking scheme, out of 40, was the following.

Introduction	4 marks
Role of person–seat models	5 marks
Assumptions behind the model	4 marks
Derivation of the equations of the model	5 marks
Expressions for T	5 marks
Complex number analysis and modulus form	8 marks
program/Algorithm	5 marks
Style and presentation	4 marks
Total	40 marks

Relevant points to make concerning this group work phase are the following.

(1) On hand to help the students were three mathematicians and between one and two physicists for the 2-hour class contact time. This is obviously costly as far as staff resources are concerned, but, having said this, all staff were kept extremely busy helping the groups.

(2) The groups, for their part, quickly realised that a division of labour was needed in order to meet the deadlines imposed. The word-smiths amongst the group concentrated on the introduction, role of person–seat models, and the assumptions behind the model. The more skillful mathematicians amongst the group worked on the derivation of the equations of motion and the expressions for T in complex number form. The algebra-crunchers did the complex-number analysis, and the computer buffs worked on designing the algorithm for BASIC program. Group reports were handed in often in three or four contrasting literary styles (you may well notice similar properties in this chapter!). It was stressed to the students that, in compiling their group reports, each group member should have appreciation of the

entire contents of the report in order to be able to undertake their individual work successfully.

(3) The reports themselves ranged from being very good to average. In general the standard of presentation and packaging of the group reports tended to lack polish. This, obviously, was due to reports being hastily written and the need to put together quickly anything up to four different sub-reports. No group, it seemed, took it upon itself to appoint a group editor to produce a more homogeneous style. Only a few groups used references to good effect, and most tended to rely heavily on the case study booklet that had been issued at the beginning.

(4) What the staff noticed at this group work stage was the emergence, either by choice or default, of group leaders who organised the group's activities. In one or two groups it was clear that some group members were doing more work than others. The students realised this, too, but no overt resentment was noticeable. Groups seemed to take notice of how other groups were getting on and this friendly competition seemed to help groups keep to their schedule. There was a great deal of groaning and complaining by the students in the first week about all the algebra that had to be done, but this was balanced by the obvious sense of achievement that a group felt when it obtained the correct expression for the real or imaginary part of the denominator of T for example. The students certainly became totally engrossed in the Case Study, and worked as teams committed to achieving their goal.

(5) It was noticeable how students used the staff resources available to them selectively. They quickly singled out, as they saw it, staff who were good at modelling, or at algebra crunching, or at program design and so on, and soon, more by accident than by design, staff were role playing. The roles staff played tended to reflect very much the roles they had played in producing the case study booklet and in working up the case study in its initial stages.

Before considering the individual work that the students undertook in the remaining two weeks of the 4-week period, it is interesting to report on how the students wanted the group work to be assessed. The students were offered two choices: the group report would be marked and each group member would receive that mark, or the group report would be marked and each group would take it upon themselves to allocate the marks amongst members as they felt was appropriate. They seemed interested in the latter idea, but opted for the former after some discussion. All the students seemed happy with the 60/40 split in marks between individual and group work assessment.

The group reports were all triple marked and handed back to the groups, with detailed comments and criticisms, at the start of the final week. This gave the students plenty of time to incorporate this feedback of information into their second and final (individual) report. This is hard work for the staff, but essential for students who obviously do not wish to repeat mistakes and who genuinely want to improve on their first mark.

4.2 Individual work

At the start of the third week the students each had an algorithm designed to evaluate T for a range of frequencies and values of the parameters M_i, S_i and δ_i, and field data on four different types of vehicle were made available. The idea was that each person in a group would work at achieving a model fit to data for a given vehicle, and thus have the opportunity of producing something individual whilst still being able to consult group members about any programming problems, for example. The fit was achieved using graphics and *seeing* what effect altering parameters had in the model. A typical example of a fit obtained is shown in Fig. 3.

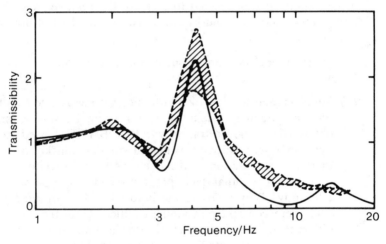

Fig. 3. Typical fit to data.

By and large this part of the case study worked very well and all students, save one, managed to achieve their own fits. The one student who did not, blatantly plagiarised someone else's results and was marked down accordingly in his final report. A summary of the marks for the individual reports is shown in Fig. 4.

The highest mark was 54/60, i.e. 90% and the lowest was 19/60, i.e. 32%. The mean mark was 41.7/60, i.e. 69.4% and the spread in marks, as measured by the standard deviation, was 7.9/60, i.e. 13.1%.

The marking scheme adopted (out of 60) was the following.

Program listing	5 marks
Display of experimental data	5 marks
Best fit to experimental data	25 marks
Discussion of methods for a systematic fit	15 marks
Presentation and style	10 marks
Total	60 marks

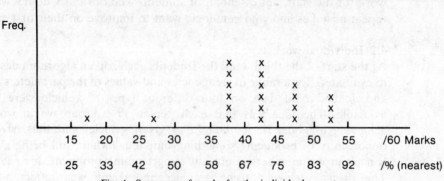

Fig. 4. Summary of marks for the individual reports.

With the students already having had the benefit of feedback on their group reports it was decided to increase the relative weighting given to presentation and style and by and large the standard of presentation did improve.

The major points that arose out of this part of the case study were as follows.

(1) Students (and staff!) took a little time getting used to (what was then) the new Econet system (how, for example, was a screen dump obtained?). Some frustration with BBC BASIC showed itself at times, but things soon settled down, and students quickly discovered why there is a copy button on the Beeb for example. Quite a few students soon found out that their programs had not been written all that efficiently. They realised, the hard way, that repeatedly evaluating what were constant polynomial coefficients, in the plotting of T for different ω values, was much slower than first evaluating these constants and repeatedly using their stored values afterwards.

(2) Most students seemed impressed by the power of computer graphics in producing a visual fit to data points. They soon became familiar with the sensitivity of T to changes in the fitting parameters, and fairly quickly were able to obtain reasonable fits. In their first year the students had seen nonlinear least squares in action, and all agreed that the visual approach of using graphics was much easier to appreciate and understand, and certainly much easier to set up.

(3) Whilst being impressed with the graphics, students and staff were not too impressed with the hardware. In the hot steamy atmosphere of the modelling sessions, the hardware seemed increasingly to crash and respond with Not Listening. Whether this was an actual overheating problem or not was never properly ascertained, but it did cause a deal of frustration, and the inevitable groans as the writing up deadlines drew nearer. In the one hour feedback session at the end we did manage to convince students that hardware problems were all part and parcel of modelling, and no less of a problem outside the Polytechnic in the real world.

(4) As in the initial stages of the case study, students again tended to use staff help selectively, and the time of those staff who were familiar with the ins and outs of the BBC micro was at a premium. As it turned out, the additional use of the Econet system in the evenings of the final week proved very popular with students, and even the HoD did a stint of supervision to help out.
(5) From the staff point of view it had been time consuming getting to grips with the new hardware and software. This coupled with the marking of the group reports, and the extra Econet supervision in the final week, gave staff something of a busy time.
(6) As for the reports submitted by the students, these were a little more polished than the group reports. The average mark was marginally up from 66% to just under 70% and some reports, despite the short time scale involved, were actually word processed and one was neatly bound. Typical reports were between ten and twenty sides of A4 and generally conveyed an air of enthusiasm for the case study.

5. ADVANTAGES AND PITFALLS OF CASE STUDIES OF THIS TYPE

What follows is a list of advantages and pitfalls.

5.1 Advantages of using case studies

(1) As well as motivating the students, by their seeing mathematics in action, case studies of this nature can provide the ideal vehicle for reinforcing and integrating mathematical and modelling ideas learnt so far. Here, for example, was a problem that brought together modelling, complex numbers, linear systems, matrices, model fitting ideas, programming, and so on.
(2) In a similar vein, case studies of this type provide an ideal opportunity for a student to feel that he has actually achieved something major using his mathematics and modelling skills. He has gone and dirtied his hands tackling a *real* problem, and he has taken the problem through from start to finish. He has seen how complex real problems can be, but he has coped with the complexity by writing his own program and using prepared software to help him. A student who achieves things becomes motivated and will, hopefully, go on to achieve greater things.
(3) In addition to developing a zeal for the subject and reinforcing mathematical ideas, case studies such as this offer the opoportunity for group work and all that this entails. Students see how important organisation, division of labour, team-work, and so on can be when faced with tackling an ambitious problem within a fairly rigid set of constraints. Case studies can certainly help to develop a student's initiative and organisational skills, and allow him to see himself in perhaps a new light within his peer group.
(4) On the student/staff front, case studies provide the ideal opportunity for staff and students to get to know one another better on a semi-formal work-centred basis.

5.2 The pitfalls

(1) It is hard work mounting a case study such as this from scratch. Real problems, being tackled at the research level, cannot always be readily brought down to a 2nd year teaching level without a great deal of careful planning beforehand. Tight marking deadlines, too, mean that staff have to be willing to put in an appreciable amount of time in making such case studies a success. There is therefore the potential pitfall of underestimating the amount of effort required.

(2) There is always the problem of trying to do too much in too short a time with the students, and instead of motivating them and developing their enthusiasm for the subject they are switched off. Certainly this case study would have been impossible in the time available had a considerable amount of new mathematics or modelling ideas been required.

(3) Finally, a student can only feel happy about what he is doing if he knows exactly what is expected of him, and how much of his efforts he needs to channel into achieving the various objectives in his remit. We feel that a case study could go seriously wrong if the student's remit (given that he is working to a tight schedule) is not spelled out in detail. We also recommend that before any report writing takes place, the students are made aware of what the likely marking scheme is to be. Failure to be precise about what exactly is required of the student can, in many ways, detract from the advantages to be gained using case studies in the tecaching of mathematics and modelling.

REFERENCES

Band, E. G. U., Payne, P. R. & Wright-Patterson, A. F. B. (1970). Ohio, AMRL-TR-70-35.
Benson, A. J., Patten, W. & Rose, C. V. (1973). *J. Cell Plast.*, **9**, 22.
Collier, P., Douglas, D., Hilyard, N. C. & McNulty, G. J. (1983). *Proceedings, INTERNOISE '83, Edinburgh*, Vol. II, 913, June 1983.
Griffin, M. J. (1978). *Applied Ergonomics*, **9**, 15.
International Standard (1981). ISO 5982.
McNulty, C. J. & Douglas, D. (1982). 4th Conference on Teaching Noise and Vibration, Sheffield City Polytechnic, 283.
Patten, W., Seefried, C. J. & Whitman, R. D. (1974). *J. Cell Plast.*, **10**, 276.
Payne, P. R. & Wright-Patterson, A. F. B. (1979). AMRL-TR-71-29.
Pollart, D. l., Seefried, C. J. & Whitman, R. D. (1974). *J. Cell Plast.*, **10**, 171.
Snowdon, J. C. (1968). *Vibration and Shock in Damped Mechanical Systems*. J. Wiley.

14

Mathematical Models in Forest Management

W. J. Reed
University of Victoria, Canada

SUMMARY

The problem of the management of a forest subject to the risk of fire gives rise to some simple models which can be profitably used in a mathematical modelling course. The problem can be modelled at two levels: (a) that of the single stand, and (b) that of the 'whole-forest'. The stand level model presented is in continuous time and provides some analytic results with interesting economic interpretations. The forest level model is in discrete-time and the results from it are numerical. The mathematical techniques used in the stand level analysis include simple probability (the Poisson process), geometric series and calculus. The forest level model is formulated using a stochastic difference equation and an approximately optimal feedback solution is found using linear programming. The results of the analyses at the two levels are complementary.

1. INTRODUCTION

A real-life problem that lends itself to being modelled mathematically in more than one way is particularly useful in the teaching of the applications of mathematics. Not only may it provide examples of the use of several different mathematical techniques, but also the results of the different analyses can be compared. Serious discrepancies in the results can point the way to model inadequacies, while complementary results from two different models will lend considerable weight to their credibility and to the credibility of the conclusions that are drawn from them.

One such area of application that I have encountered in my own research concerns the problem of assessing the effects of forest fire on forest yield and forest harvesting policies. The problem can be modelled at two levels: (a) that of the single stand, and (b) that of the 'whole-forest' comprising many stands of different ages. I have found it convenient to model the single stand in continuous time and the whole-forest in discrete time. In teaching, this provides a useful illustration of how a discrete-time formulation is more convenient in some instances and a continuous-time formulation more convenient in others.

The mathematical techniques involved in the two levels of modelling are quite distinct. In the stand-level model the techniques associated with geometric series and simple probability theory (the Poisson process) are used. In the forest-level model dynamic equations for the evolution of the forest subject to fire, are used. These are formulated in matrix terms. The notion of feedback control for stochastic dynamic systems arises, and a procedure for determining approximately optimal harvests can be found through the use of linear programming.

The economic phenomenon of time discounting occurs in both models; in continuous-time form for the stand-level model and in discrete-time form for the forest-level model. While mathematical analysis of the stand-level model leads to some general conclusions with economic interpretations, the results of the forest-level analysis are essentially numerical. Pedagogically this provides a useful illustration of the different kinds of output that can be expected from mathematical models. It turns out that the results of the two levels of modelling are complementary in the areas where one would expect.

In the next two sections a brief description of the models and some results are given. For the sake of brevity much of the mathematical detail is omitted. Emphasis is placed on the aspects of the models which are of pedagogic interest.

2. STAND-LEVEL ANALYSIS

Suppose that a stand of trees of age a growing on a site has a net *stumpage value* (value of timber net of cutting and transportation costs) of $V(a)$. Suppose further that a stand is cut whenever it reaches some *rotation age*, T, and that the site is then subsequently replanted with trees of similar growth characteristics. If costs of clearing and replanting the site after a cut are c_1, then the total present (discounted) value of the stream of revenues (and costs) from the site (starting with a newly planted site) is

$$\sum_{n=1}^{\infty} e^{-n\delta T}(V(T) - c_1) \tag{1}$$

where δ is the instantaneous discount rate related to the per annum discount rate i, by $\delta = \ln(1 + i)$.[1] The above present value is known traditionally as the *land expectation value* (LEV) (see e.g. Clark, 1976, p. 259). The expression (1) is a geometric series and can be summed to give

$$\text{LEV} = \frac{V(T) - c_1}{e^{\delta T} - 1} \tag{2}$$

[1] In the classroom some discussion of present value, time discounting and the relationship between the appropriate rate for annual compounding and instantaneous compounding (such as that in Clark, 1976, p. 69) might be necessary at this point.

The optimal rotation age can be determined by setting the derivative of (2) equal to zero. Thus the optimal rotation age T^* solves

$$V'(T) = \frac{\delta(V(T) - c_1)}{1 - e^{-\delta T}} \tag{3}$$

This result is known as the Faustmann formula (Faustmann, 1849). An economic interpretation of the result can be obtained by multiplying both sides of (3) by an infinitesimal time h and re-expressing it as

$$V'(T)h = \delta h(V(T) - c_1) + \delta h \sum_{n=1}^{\infty} e^{-n\delta T}(V(T) - c_1) \tag{4}$$

again using the formula for the sum of a geometric series. The term on the left represents the growth in the value of a stand of age T during the infinitesimal time h if cutting does not take place. The first term on the right represents the growth in value of the revenue earned (interest) through cutting the stand at age T, while the second term (which is $\delta h \times$ (LEV)) is the incremental growth in the value of the site (i.e. the interest that could be earned on revenue obtained through selling the site). Thus, *optimally, one sets the rotation age to the age where the growth in value of the stand through not cutting equals the growth in the value of revenue that could be earned through cutting the stand and selling the site.* It should be pointed out in the classroom that results such as this, where marginal growth rates are equal at the optimum are common in the models arising in microeconomic theory.

Suppose now that stands are destroyed from time to time by fire. Suppose that subsequent to a fire, costs c_2 of clearing and replanting the site are incurred, and that after replanting, a new stand will grow. A typical evolution of site might look like Fig. 1. We need some probabilistic model for the occurrence of fires. The simplest possible model is that fires occur independently of one another (in time) and at random (in time), i.e. that fires occur in a Poisson process (see e.g. Devore, 1982, p. 118). If we denote the times between successive destructions of the stand (either by fire or by cutting) by $X_1, X_2, \ldots$, then the X_i's are independently identically distributed random variables with cumulative distribution function (c.d.f.)[2] given by

$$F(x) = \text{pr}(X \leq x) = \begin{cases} 1 - e^{-\lambda x}, & x < T \\ 1 & x \geq T \end{cases} \tag{5}$$

(where λ is the average rate of fires in the Poisson process), i.e. the times between successive destructions have an exponential distribution, truncated at $X = T$ and with an atom of probability of size $e^{-\lambda T}$ at $X = T$.

---

[2]Most students with an introductory course in Probability or Statistics should be familiar with the notion of a c.d.f. and the fact that the interarrival times in a Poisson process have an exponential distribution. If not the derivation from the Poisson postulates provides a good example of the use of the c.d.f. (see e.g. Devore, 1982, p. 154).

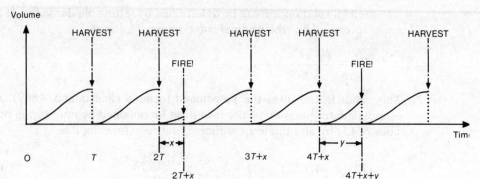

Fig. 1. Possible evolution of a site. In rotations 1, 2, 4, 5 and 7 the stand grows until the cutting age T. In rotation 3 it is destroyed by fire at age x and no harvest is realised. Similarly in rotation 6 it is destroyed by fire at age y.

Associated with each destruction of a stand there is a revenue, Y. For the nth destruction the revenue is

$$Y_n = \begin{cases} -c_2 & \text{if } X_n < T \text{ (fire)} \\ V(T) - c_1 & \text{if } X_n = T \text{ (harvest)} \end{cases} \tag{6}$$

The *expected* (discounted) present value of the random stream of revenues from the site (the land expectation value) is

$$\text{LEV} = E\left\{ \sum_{n=1}^{\infty} e^{-\delta(X_1 + X_1 + \ldots + X_n)} Y_n \right\} \tag{7}$$

Since Y_n is independent of $X_1, \ldots, X_{n-1}$ (but not of X_n), (7) can be written as

$$\text{LEV} = \sum_{n=1}^{\infty} E(e^{-\delta(X_1 + \ldots + X_{n-1})}) E(e^{-\delta X_n} Y_n) \tag{8}$$

After some integration and again using the formula for the sum of a geometric series, (8) can be summed to give:

$$\text{LEV} = \frac{(\lambda + \delta)(V(T) - c_1)e^{-(\lambda+\delta)T}}{\delta(1 - e^{-(\lambda+\delta)T})} - \frac{\lambda}{\delta} c_2 \tag{9}$$

(See Reed, 1984 for details.) The optimal cutting are T^* can be obtained by setting the derivative of (9) equal to zero. That is T^* satisfies

$$V'(T) = \frac{(\lambda + \delta)(V(T) - c_1)}{1 - e^{-(\lambda+\delta)T}} \tag{10}$$

This is of the same form as the Faustmann equation (3), with the discount rate δ replaced by $\lambda + \delta$. From this it can be seen that *the effect on the rotation period of a risk of destruction by fire is the same as that of adding a premium to the discount rate of an amount equal to the average rate at which fires occur.*

An economic interpretation to (10) similar to that in the no-fire case can be given. Multiplying both sides by an infinitesimal time h, (10) can be expressed as:

$$V'(T)h(1 - \lambda h) + [-V(T) + c_1 - c_2]\lambda h \\ = \delta h[V(T) - c_1] + \delta h\, \text{LEV}(T) + o(h) \qquad (11)$$

The r.h.s. is the same as the r.h.s. of (4) and represents the growth in revenue (interest) that could be earned, in the time increment h, through cutting the stand and selling the site, while the l.h.s. represents the *expected* growth in value of the stand given that there is no cut. To see this the l.h.s. can be written in terms of conditional expectation as

E(growth in value of stand with no cut/no fire) pr(no fire)
$+ E$(growth in value of stand with no cut/fire) pr(fire)

since in an infinitesimal time increment of length h the probability of a fire (an event in a Poisson process) is λh.

The results above are all theoretical in nature and relate only to the optimal rotation age T^*. To determine the effect on land expectation value of the presence of the risk of fire, numerical methods must be used, since in general solution to (10) can only be carried out numerically. A suitable classroom approach is to plot the LEV (10) as a function of T for various rates of fire λ. An example is shown in Fig. 2. The growth curve used was that for a hectare of spruce growing in the Fort Nelson region of NE British Columbia (see Reed & Errico, 1985a, for details) and the per-annum discount rate was 3%. It can be seen how an increase in the rate of fires λ causes a small reduction in the optimal rotation age T^*, but a considerable reduction in the land expectation value. In particular with a fire rate of $\lambda = 0.005$ (on average one fire every 200 years) the LEV is reduced to approximately 50% of its value with no fires present. For fire rates in excess of $\lambda = 0.0104$ (one fire every 96 years) no net rent can be extracted from the resource in the long-run. The historical rate of fires in the region is estimated to be about $\lambda = 0.013$.

Of course the results above depend on the parameter values used. A useful homework exercise is to ask students to repeat the analysis for different values of the discount rate parameter δ and the cost parameters c_1 and c_2, and to determine the sensitivity of the results to these various parameters.

Another useful exercise is to ask students to criticise the model in terms of the realism of its assumptions and other shortcomings, such as important aspects omitted. Some of the things that they might come up with, such as age-dependent fire probabilities, the possibility of partial salvage after a fire etc., can quite easily be handled (see Reed & Errico, 1985a). Other difficulties such as uncertainties over future stumpage values (dependent on price and demand for lumber), uncertainties (statistical and otherwise) over the growth characteristics of current and future stands and uncertainties over future fire probabilities etc., can less easily be handled

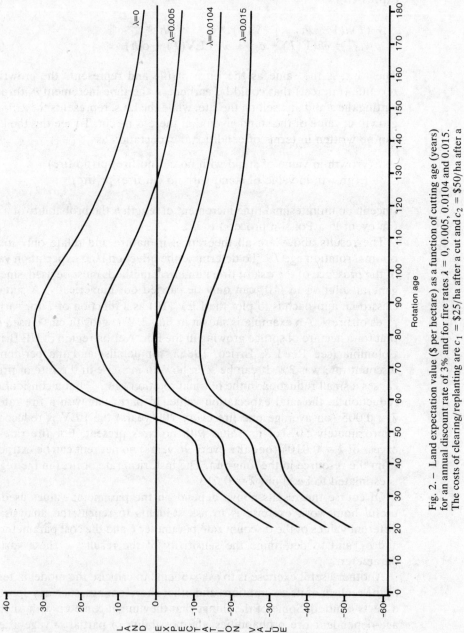

Fig. 2. – Land expectation value ($ per hectare) as a function of cutting age (years) for an annual discount rate of 3% and for fire rates $\lambda = 0, 0.005, 0.0104$ and 0.015. The costs of clearing/replanting are $c_1 = \$25$/ha after a cut and $c_2 = \$50$/ha after a fire.

mathematically. This could lead to a useful discussion on the role and limitations of mathematical modelling in general.

If one is very lucky a student might criticise the model along the following lines:

> A forest or timber supply area for a mill centre will typically comprise many stands. If each stand in the forest is independently managed in an optimal fashion, the flow of timber to the mill will likely be very erratic, especially in Western North America, where there is much old growth timber still to be cut. Clearly a more stable flow of timber would be preferred. How can one incorporate this into the model?

The answer to such a fortuitous question would be that one needs a forest-level model to accommodate this aspect. In the next section, such a model, formulated in discrete time is discussed.

3. A FOREST LEVEL MODEL AND ANALYSIS

Let the vector $\mathbf{x}_t = \{x_1^t, x_2^t, \ldots, x_k^t\}'$ denote the areas in the forest with trees in age classes $1, 2, \ldots, k$ at the start of period t, and let the vector $\mathbf{h}_t = \{h_1^t, h_2^t, \ldots, h_k^t\}'$ denote the areas harvested in each of the age-classes $1, 2, \ldots, k$ in period t, Let $\mathbf{v}' = (v_1, v_2, \ldots, v_k)$ denote the average volume per hectare of stands in age-classes $1, 2, \ldots, k$. Then the total volume harvested in period t will be $H_t = \mathbf{v}'\mathbf{h}_t$.

Suppose now that $\theta_1^t, \theta_2^t, \ldots, \theta_k^t$ are *random variables* representing the *proportions* of the areas destroyed by fire in age-classes $1, 2, \ldots, k$ in period t. A dynamic model for the evolution of the forest is given by the *stochastic difference equation*

$$\mathbf{x}_{t+1} = R_t \mathbf{x}_t - S_t \mathbf{h}_t \tag{12}$$

where R_t and S_t are random matrices:

$$R_t = \begin{bmatrix} \theta_1^t & \theta_2^t & & \theta_k^t \\ 1 - \theta_1^t & & & \\ & 1 - \theta_2^t & & \\ & & 1 - \theta_{k-1}^t & 1 - \theta_k^t \end{bmatrix} \tag{13}$$

$$S_t = \begin{bmatrix} -1 + \theta_1^t & -1 + \theta_2^t & & -1 + \theta_k^t \\ 1 - \theta_1^t & & & \\ & 1 - \theta_2^t & & \\ & & 1 - \theta_{k-1}^t & 1 - \theta_k^t \end{bmatrix}$$

The appropriateness of this model can be checked by multiplying out the r.h.s. of (12). It will be seen that the area, x_{t+1}^1, in age-class 1 at the start of period $t + 1$ comprises those areas harvested in period t, plus the parts of those areas not harvested, that are destroyed by fire. The area in age-class

2 at the start of period $t + 1$ is the area formerly in age-class 1 which was not harvested nor burnt. Note that it is assumed that age-class k comprises all stands of age greater than k periods.

The expected (discounted) present value of the stream of volumes harvested is

$$J = E\left\{\sum_{t=1}^{\infty} \alpha^t \mathbf{v}' \mathbf{h}_t\right\} \tag{14}$$

where α is the per period discount factor, and is related to the per annum discount rate by $\alpha = (1+i)^{-a}$, where a is the length in years of a period.

To ensure a degree of evenness in the flow of timber from the forest, harvest flow constraints could be imposed. For example, we might have constraints

$$(1 - \gamma_1)\mathbf{v}' h_{t-1} \leq \mathbf{v}' \mathbf{h}_t \leq (1 + \gamma_2)\mathbf{v}' h_{t-1} \quad t = 2, 3, \ldots \tag{16}$$

which would ensure that the percentage change in volume harvested from period to period would lie within specified bounds.

We could look now for an optimal policy to maximise (14) subject (12), (16) and constraints of the form

$$\mathbf{O} \leq \mathbf{h}_t \leq \mathbf{x}_t, \quad t = 1, 2, \ldots \tag{17}$$

This is a problem in stochastic control since the dynamic equation (16) is stochastic. For students with some familiarity with control theory the difficulties of finding solutions to stochastic control problems could be discussed at this point. For example, the use of dynamic programming and its limitation because of the problem of dimensionality could be discussed. In any case it should be pointed out that the optimal policy for this stochastic control problem will be of a *feedback* nature. That is that the optimal harvest $\mathbf{h}_t^*$ in period t cannot be determined prior to period t, but will depend on the current state $\mathbf{x}_t$ and on previous harvests and states (possibly).

Exact solution to the control problem cannot be obtained but an approximately optimal solution can be found using the principle of *certainty equivalence* (see e.g. Chow, 1975). To do this one replaces the stochastic equation (12) by the *deterministic* equation

$$\mathbf{x}_{t+1} = \bar{R}\mathbf{x}_t - \bar{S}\mathbf{h}_t \tag{18}$$

where $\bar{R}$ and $\bar{S}$ are the expected values of the random matrices R_t and S_t, and then solves the resulting *deterministic* control problem. The optimal first period harvest for this problem is determined and after this harvest has been made, and the random fires have occurred, the new state $\mathbf{x}_2$ is observed. One then repeats the above procedure using $\mathbf{x}_2$ as the initial state vector to determine the harvest $\mathbf{h}_2$. One then continues to iterate the procedure solving a new deterministic control problem at each step (see Reed & Errico, 1985b for details).

The deterministic control problems that have to be solved are of the form, maximise (14) subject to (16), (17) and (18). To find a solution one can first replace the objective (14) by one with a finite time horizon

$$J' = \sum_{t=1}^{N} \alpha^t v' \mathbf{h}_t + \alpha^{N+1} \mathbf{r}' \mathbf{X}_{N+1} \tag{19}$$

where $r' = (r_1, r_2, \ldots, r_k)$ is a vector of expected present values of single hectares of forest in age-classes $1, 2, \ldots, k$ (determined from a single stand model).

The problem of maximising (19) subject to (16) (17) and (18) is linear (in the $\mathbf{x}_t$ and $\mathbf{h}_t$) in both its objective and constraints and thus can be solved by *linear programming* (LP) using for example the Revised Simplex Algorithm (see e.g Childress, 1974). As an exercise students can be asked to write out the tableau for the LP problem.

In Fig. 3 paths (a) and (b) show the sequence of values of $H_t = \mathbf{v}' \mathbf{h}_t$ for two such solutions corresponding to a zero fire probability (a) and a 1% per

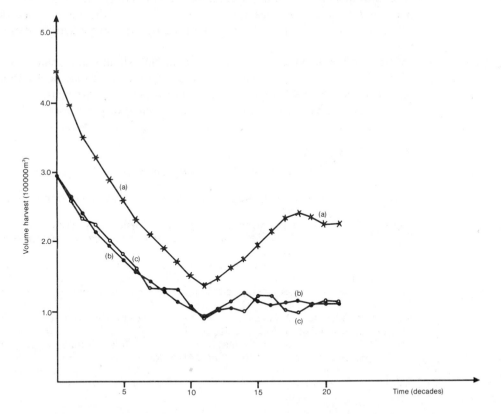

Fig. 3. Volumes harvested over time with a per annum discount rate of 3% Path (a) shows the optimal harvest sequence when there is no risk of fire. Path (b) show the predicted harvest sequence when there is an (age-dependent) probability of fire of 1% per annum. Path (c) shows one 'actual' sequence of harvests using the certainty equivalence procedure, discussed in the text and using a random number generator to generate fires with a 1% per annum probability.

annum fire probability (b). The latter path can be regarded as an *estimate* or *prediction* of future volumes harvested under optimal management if in fact a 1% fire probability prevails. Path (c) shows an 'actual' sequence of harvests obtained using the certainty equivalence procedure above, and using a random number generator to simulate fires with a 1% per annum probability of occurrence. Although this is only one possible sample path of many, it can be seen that in this case the solution to the deterministic problem (path (b)) provides a good prediction of future harvests.

Numerical comparisons between the results of the stand-level and forest-level analyses can be made. In the forest-level model, in steady-state (after about twenty decades in the example) the harvest flow constraints are non-binding and each stand is effectively managed independently. The long-run average yield per annum per hectare can be compared with that for a single stand model. A very close agreement has been found (see Reed & Errico, 1985 (b)) the only differences being due to the different (discrete/continuous) time formulations. On examining the harvest schedules in steady-state in the forest-level model one finds that areas are harvested as soon as trees reach a given age. To the nearest decade (discrete time unit) this age agrees with the optimal single stand rotation age.

These points of agreement between the models should be emphasised in the classroom. While they do not guarantee that the models are an accurate representation of reality, they do offer reassurance that the mathematical techniques of analysis, and the subsequent calculations, are correct.

REFERENCES

Chow, A. C. (1975). *Analysis and Control of Dynamic Economic Systems*. J. Wiley, New York.
Childress, R. D. (1974). *Sets, Matrices and Linear Programming*. Prentice-Hall, Englewood Cliffs, NJ.
Clark, C. W. (1976). *Mathematical Bioeconomics*. J. Wiley, New York.
Devore, J. L. (1982). *Probability and Statistics for Engineering and the Sciences*. Brooks/Cole, Montery, CA.
Faustmann, M. (1849). Berechnung des Werthes, welchen Walboden sowie nach nicht haubare Holzbestande. *Allgemeine Forst und Jagd Zeitung*, **25**, 441–445.
Reed, W. J. (1984). The effects of the risk of fire on the optimal rotation of a forest. *J. Environ. Econ. Manag.*, **11**, 180–190.
Reed, W. J. & Errico, D. (1985a). Assessing the long-run yield of a forest stand subject to the risk of fire. *Can. J. For. Res*.
Reed, W. J. & Errico, D. (1985b). Optimal harvest scheduling at the forest level in the presence of the risk of fire, *Can. J. For. Res*.

15

Case Studies and CAL in Engineering Mathematics

M. Stewart Townend
Liverpool Polytechnic, UK

SUMMARY

Various recent reports, for example, Lighthill (1979), Finniston (1980) and Cockcroft (1982), stress the need for mathematics to be presented to students in a way which they perceive to be relevant.

This chapter describes an attempt to achieve this for mechanical engineering students through the use of engineering related case studies, CAL packages and numerical software libraries.

The advantages, disadvantages and implications of such a teaching technique are discussed. A suite of successfully used examples is appended.

1. INTRODUCTION

> Modern Engineering needs Mathematics
> What sort of Mathematics?

Applied Mathematics—Applied in the sense of how mathematical formulation of physical laws can be effectively applied to practical problems.

Teaching must be constantly permeated with concrete examples (enhances motivation).

We must communicate the techniques used to represent an engineering system in mathematical terms.

The solution of the problem itself is not a difficulty due to computational facilities.

These ideas can be best communicated through project work involving 'real world' problems.

J. Lighthill, Bulletin IMA, April, 1979.

After their own engineering specialism, mathematical ability is the next most important skill required by engineers.

Despite this, it is a fair generalisation to say that prior to the information technology explosion the usual mathematics course offered to undergraduate engineers consisted of a catalogue of analytical techniques which provided closed form solutions to problems which had been deliberately contrived to possess closed form solutions. Such was the level of simplification used in some of these examples that even if the original problem were of engineering interest, the solution bore little or no relation to it!

As Lighthill (1979) observed, we must teach our students to apply mathematics effectively to their problems. To do this requires more than skill in 'technique bashing' (although this has its place); it requires the ability to formulate the problems mathematically, solve the mathematical problem and then assess the value of the solution. In short it requires the skills of mathematical modelling.

Of course our engineering colleagues develop their students' engineering modelling skills within the parent department. This background can be exploited in the mathematics courses through the modelling and solution of further engineering problems.

Due to the rate of change of technology which today's students will experience during their professional career, very few of the methods which we teach them will still be current in ten or twenty years time. It is thus paramount that our teaching should go beyond mere techniques, and should additionally impart to the students some idea of the thinking behind the mathematical formulation of their engineering problems. We must teach them to teach themselves, for the real world problems which they may encounter will not always match neatly one of their undergraduate exercises!

Matters are further complicated by the fact that currently one is trying to attain the objectives while preparing the students for a conventional written examination of a traditional mathematical methods syllabus. Based on my experiences with the first and second year of a BSc. Mechanical Engineering course I am confident that all these objectives are attained through the use of case studies and CAL sessions to augment formal lectures and tutorials.

2. USE OF CASE STUDIES

The reason for using a tutor-driven case-studies approach, to illustrate the mathematical methods contained in the syllabus, rather than a student-driven modelling approach is the logistical and resource constraints. The class upon which this report is based contains 55 students and, even with some tutorial assistance from two colleagues, a modelling approach is not considered feasible given the need concurrently to prepare the students for a conventional written examination.

The students are encouraged to participate in the case-study sessions as much as possible, and I act as arbiter and secretary. The major difference between this aprproach and that of mathematical modelling is that I, at least, have a clear idea of the likely form of the resulting problem formulation. Indeed some of the case studies are designed to lead to a particular formulation as a *raison d'être* for the inclusion of a specific topic in the mathematics syllabus.

I admit that this is not the best way to proceed but given the class size and the current timetable and syllabus constraints it represents a reasonable compromise.

Appendix I contains a suite of case studies which have been used with the students, together with an indication of the mathematics required. Student reaction has been favourable with much comment about the extent to which the mathematics course has been integrated with their engineering studies; this has enhanced their motivation in both their engineering and mathematics courses. Case studies such as the vibration problems contained in Appendix I lend themselves quite naturally to the idea of a sensitivity analysis which can rapidly be performed using some of the numerical software which is available to the class.

The product of this teaching approach is a student who has some skill in formulating engineering problems in mathematical terms, a range of techniques for their solution, and an awareness of the need to assess the quality of the solution. During the academic year, the students are required as part of the course-work assignments to produce a written report on one or two case studies, in the same style as a laboratory report, in order to develop their communication skills—often a serious weakness in undergraduates, as reported by McLone (1973). Looking ahead, I hope that the benefits of this teaching approach will be recognised by the parent department, thus permitting a change to a modelling approach linked even more closely with the students' engineering studies. Such recognition will require additional timetabled hours or a reduction in content of the mathematical methods course together with additional staff assistance.

3. USE OF CAL AND SOFTWARE LIBRARIES

Concurrent with the development of case studies and modelling exercises for teaching mathematics to undergraduates has been a dramatic increase in the availability of interactive computing facilities, good software libraries such as NAG (see Appendix II), and CAL packages such as CALNAPS (see Appendix II). These are fully exploited in the case studies examined and enable the students to progress from a possibly unrealistic, closed form, solution to obtain more realistic solutions without the tedium of laborious algebra or repetitive numerical calculations. Since the results are relatively easily obtained via the computational facilities, it is reasonable to expect the student to address such additional questions as

— how sensitive are the results to changes in the parameters or conditions?
— is the solution stable?
— is it sensible?
— is it accurate?
— do I believe it?

Attention to questions such as these is considered to add an important extra note of reality to the exercises.

The CAL package used with our engineering students is called CALNAPS and was originally developed at Kingston Polytechnic (see Appendix II). It consists of three programs which can be used for the teaching/learning of topics in engineering mathematics such as

— solution of linear simultaneous algebraic equations,
— numerical integration,
— solution of initial value problems.

The package provides the user with the opportunity to select both a problem and a method of solution from two menus; alternatively the user can specify his own problem. Once the solution has been obtained, the user can select the method of presentation of the results (either tabular or graphical) and also the content of the results (solution values, absolute or relative errors), and so on. The package is user friendly, with many opportunities built in to access a helpfile for additional information and guidance.

Once the students have been introduced to a numerical topic in their lectures, the package has then been used successfully to illustrate various numerical properties associated with that topic. For example, the class may have had some introductory lectures on the numerical solution of initial value problems using elementary methods such as Euler and second order Runge–Kutta. Time constraints restrict the students' hand calculated efforts to the development of only a couple of steps of the solution, using one or two different step sizes. The students are then introduced to CALNAPS, and use it interactively to experiment with a range of methods and/or step sizes applied to a variety of initial value problems. In this way they quickly develop a feel for the significance of the order of a method, the sensitivity of the solution to the step size, and the concepts of stability and convergence.

The other programs in the package can be used similarly; for example, the algebraic equations package can be used to demonstrate the virtue of a pivoting strategy (the package permits inspection of each stage of the elimination process) and the concept of ill condition.

The students can obtain a print out of their terminal session, which is used to support the analysis of the methods in subsequent lectures.

Some of the case studies presented in the course can be formulated as a mathematical problem with one of the areas covered by the CALNAPS package. Once the students have become familiar with the workings of the package, which experience has shown only takes about one hour, they can use it to generate their numerical solution to the case study. Not all the case studies fit into such a small sub-section of undergraduate numerical methods, and there is a need for expertise in the use of wider ranging and more robust software; the material currently used for this purpose is the NAG library. This was chosen since the routines contained therein are well documented, efficient and representative of some of the numerical techniques currently used in industry. Their use thus provides the students with experience likely to be of benefit in their subsequent professional careers.

The NAG library is divided into chapters according to the nature of the problem to be solved (e.g. quadrature, interpolation, roots of one or more transcendental equations, ordinary differential equations). Each chapter follows a similar pattern of introduction, a list of available routines, selection of a routine for the user's problem, and documentation of each (including an example coded in both FORTRAN and ALGOL). Consequently, only one chapter is discussed in detail, and the students are taught how to use the mainframe computer to locate systematically a routine suitable for their particular problem by responding to an intelligent database with keywords.

Appendix III contains a suite of exercises and case studies which have been used in conjunction with CALNAPS and NAG.

4. REACTIONS

Student reaction to CALNAPS and NAG has been very favourable. CALNAPS has proved especially popular with the students, as they have been able to see for themselves the answers to the 'What if...?' type of questions they have askd in class, without the need to spend several hours with a calculator. The scope and power of the NAG library has also made a considerable impression, clouded only by the difficulties which some students encountered in interpreting the generalised FORTRAN coding of the NAG manual. Nevertheless several student groups have requested 'more NAG and earlier in the course' in order that they can use it more extensively in connection with their engineering studies.

5. CONCLUSIONS

In addition to the conventional lecture and tutorial teaching technique, case studies are integrated into the timetable in order to relate the contents of the mathematics syllabus to the mainstream engineering studies through the solution of a variety of engineering problems. Experience has shown that this increases motivation.

The range of case studies presented is sufficient to communicate the techniques used to represent engineering problems mathematically.

CAL packages are used to help students develop a feel for the strengths and weaknesses of the different numerical methods included in their mathematics syllabus while software libraries such as NAG give access to robust software for obtaining the solutions to numerically posed case studies.

The mix described has been demonstrably successful as evidenced by the students' motivation and written reports. Further development is dependent upon the allocation of extra timetabled hours and additional staff assistance to permit a change to a student-driven modelling approach.

APPENDIX I: CASE STUDIES IN ENGINEERING MATHEMATICS

Vibration absorption The damping of the vibrations of the car deck of a car ferry and discussion of the difficulties associated with the variation of frequency of the vibrations due to variations in ship's speed.

The problem is modelled as a coupled masses/springs problem leading to simultaneous ordinary differential equations and solution via harmonic trial solutions. (Ref. OU Course MST 204.)

Opening and closing of a swing door Modelled as damped/rotational system with a restoring force.

Analysis of the solution for different damping and spring constants leading to discussion of the practical importance of critical damping.

Computer disc drives Typical data for the seek time for a computer disc drive are

minimum seek time
maximum seek time
average seek time

Students often query why the average seek time is not the mean of 25 ms and 60 ms.

Modelling of this problem leads to a probabilistic problem. (Ref. A. E. Hart, On the mean—a study in modelling, *Teaching Math. and its App.*, Vol. 1, No. 2, 1982.)

Oar arrangements in rowing Presentation of some unconventional (but IOC approved) oar arrangements which an elementary mechanical analysis reveals have some theoretical advantages over more conventional rigs. (Ref. M. S. Thompson, *The Mathematics of Sport*, Ellis Horwood (1984).)

Tumble drier design In a tumble drier, the clothes spend part of the time travelling on the wall of the drier and part of the time falling back through the air towards the bottom.

What angular velocity should a tumble drier have in order to dry the clothes as quickly as possible?

How does your result compare with the angular velocity of commercial tumble driers?

Modelling of this problem involves motion in a vertical circle and the determination of extreme values. (Source: OU Summer School, 1984.)

Wheel balancing Small capacity motorcycles are fitted with fairly narrow wheels and tyres. In order to balance such wheels it is adequate to attach a small weight to the lightest side of the wheel. This is called static balancing.

Some idea of the need for wheel balancing can be gained by considering a machine, travelling at 60 mph, for which the rolling radius of the wheel is 11" and the weight required to bring the wheel into balance is 20 g (an appalling mix of units although they *are* the units used in the wheel and tyre trades!).

The current range of superbikes has wheels and tyres which are up to 5" wide. Explain why static balancing of such wheels is inadequate and consequently has led to the technique of dynamic balancing.

By taking appropriate measurements on a superbike of your choice, estimate the force acting on each fork leg if the balance weight is situated slightly off the centre-line of the wheel. What would be a reasonable value of this offset? What effects would these have on the wheel and hence on the motorcycle? This study requires the ideas of centrifugal force and moment of a force. (Ref. *Motorcycle Sport*, July, 1984.)

Spin in ball games Many ball games involve an impact between a moving ball and a fixed plane surface (e.g. basketball throws against the backboard, snooker shots off the cushion etc.).

Elementary analysis of such impacts is usually based upon a purely translational motion of the ball, and the principles of conservation of momentum and energy.

This is inadequate, since the ball almost certainly has a rotational motion in addition to its translational motion.

How does allowance for such spin effects alter the motion of the ball after impact with the plane surface?

Biomechanics of place kicking A set of numerical data representing the position coordinates of the hip, knee and ankle of a rugby player's leg during the short time-intervals immediately before and after a place kick is presented to the class.

The objective is to obtain estimates of the velocity and acceleration of the three joints.

This study requires the use of finite difference approximations; the acceleration values illustrate the need for smoothing routines. (Ref. S. Townend, 'Getting a kick out of numerical differentiation', *Teaching Mathematics and its Applications*.)

APPENDIX II

CALNAPS A computer-aided learning package developed by D. Katsifli and D. J. Fyfe of

School of Mathematics
Kingston Polytechnic

Penrhyn Road
Kingston Upon Thames KT1 32EE,
UK

NAG An extensive library of software packages covering a very broad spectrum of numerical methods. Developed by

Numerical Algorithms Group Ltd
NAG Central Office
Mayfield House
256 Banbury Road, Oxford, OX2 7DE
UK

APPENDIX III: ENGINEERING MATHEMATICS EXAMPLES USING CALNAPS OR NAG

A. Pin-jointed frames
In the analysis of pin-jointed frames, the following system of equations was obtained for the displacements U_i, V_i, in the system

$$\begin{bmatrix} 1.125 & 0 & 0 & -0.217 & -0.125 \\ 0 & 0.952 & -0.217 & 0 & 0 \\ 0 & -0.217 & 1.125 & 0 & -1 \\ -0.217 & 0 & 0 & 0.952 & 0.217 \\ -0.125 & 0 & -1 & 0.217 & 1.125 \end{bmatrix} \cdot \begin{bmatrix} V_1 \\ U_2 \\ V_2 \\ U_4 \\ V_4 \end{bmatrix} = \begin{bmatrix} 0 \\ 0 \\ -2.40 \\ 0 \\ 0 \end{bmatrix}$$

(Source: Mechanical Engineeing Undergraduate Project.)

Using CALNAPS,
 (i) Investigate the solution of the system by comparing the Gaussian elimination methods of solution.
 (ii) Use Jacobi and Gauss Seidel iteration to try and obtain the solution
   (a) starting with any initial guess at the solution,
   (b) starting with the solution obtained by any of the methods in (i).

Tutor notes: (i) illustrate the virtues of pivoting,
               (ii) enables comparisons to be made concerning convergence performance.

B. Aerodynamic forces
In a problem involving the drag and lift of aeroplane wings the following system of equations results

The right-hand sises were obtained experimentally and are rounded to two decimal places. Correct to three decimal places the values are 0.951, 0.669 and 0.521. (Source: L. Fox, *NAG Newsletter* 2/1983.)

Use the CALNAPS package to compute solutions for both sets of right-hand sides. Comement on your results. (Represent $\frac{1}{3}$, $\frac{1}{6}$ to the same accuracy as the right-hand sides.)

Tutor notes: A practical problem to demonstrate ill condition.

C. Crankshaft torque

The crankshaft of a diesel engine rotates at R radians per second. The torque applied to the crankshaft, called the throttle torque, is of the form kx, where x is the displacement of the throttle position and k is a constant. If I is the moment of inertia of the crankshaft, the inertial torque acting against the throttle torque is $I(dR/dx)$. A damping torque of the form cR, c a constant, opposes the throttle torque.

$$I\frac{dR}{dx} + cR = kx.$$

Let $I = 1$, $c = k = 0.8$ and $R = 0$ when $x = 0$.

(i) Solve the differential equation analytically. [5 marks]
(ii) Use the CALNAPS package to solve the differential equation numerically using

(a) Euler's Method,
(b) Runge–Kutta order 2,
(c) Runge–Kutta order 4,
(d) Milne–Simpson predictor corrector.

Compute the solution as far as $x = 3$ and take $h = 0.1$ and 0.5 in each case. Compare your numerical results with your analytical solution.
[10 marks]

(iii) Write a brief report summarising your results, indicating which is the best (or worst) numerical method in this problem, what effect the change in step length makes, and any other comments you feel are appropriate. [10 marks]

D. Nonlinear damping

(a) 'In the past engineers have been restricted by being unable to analyse, or to optimise, the performance of their designs due to the impracticability of solving realistic mathematical models.'

Discuss this statement briefly and explain how the use of libraries of numerical software such as NAG free designers from some of these restrictions.

(b) The equation below is known as Rayleigh's equation and occurs frequently in the dynamic analysis of mechanical systems

$$\ddot{x} + a\dot{x} + b\dot{x}^3 + cx = 0; a, b, c \text{ constants}.$$

A typical set of initial conditions might be $x = 0$, $\dot{x} = 0.5$ at $t = 0$. Using the NAG document provided (a copy of routine DØ2BBF) outline how a computer solution to this equation may be achieved for $0 \leq t \leq 1.0$. (Give your answer in outline employing pseudo-code.) (Source: Liverpool Polytechnic Examination Question.)

E. Cantilever deflections

The deflection y measured at various distances x from one end of a cantilever is given by

$2x$	0.00	0.2	0.4	0.6	0.8	1.0
y	0.0000	0.0347	0.1173	0.2160	0.2987	0.3333

Estimate the value of the deflection when $x = 0.5$.

(a) using the data values for $0.2 \leq x \leq 0.8$,
(b) using all the data.

(i) By any available means, find details of a NAG routine which will solve this problem to an appropriate accuracy. [5 marks]
(ii) Write and run a FORTRAN program, using your chosen NAG routine, to solve the problem. [10 marks]
(iii) Write a brief report, summarising your results, indicating which is the best (or worst) numerical method in this problem, what effect the change in step length makes, and any other comments you feel are appropriate. [10 marks]

Tutor notes: (i) A straightforward interpolation problem.
(ii) (a) and (b) above demonstrate the effect of including additional data.

F. Cam design 1

The velocity v mm/s of a point on an eccentric cam at a certain instant is given by

$$v = \frac{x}{3} - \ln x,$$

where x is the displacement (mm).

For what value of x is the velocity zero? Give your result correct to four significant figures.

(i) By any available means, find details of a NAG routine suitable for the solution of this type of problem. [5 marks]
(ii) Write and run a FORTRAN program, using your chosen NAG routine, to solve the above problem. [10 marks]
(iii) Write a brief report explaining how you arrived at your choice of NAG routine, what the routine needs to use it, a print out of your program and the final results. [5 marks]

Tutor notes: (i) A straightforward iterative solution of $f(x) = 0$.
(ii) There are, in fact, two solutions; experience shows that many students overlook the second.

G. Cam design 2

The non-dimensionalised displacement output of a cam y ($y \in [0,1]$) can be represented as a polynomial in x, the non-dimensional input angle ($x \in [0,1]$), as

$$y = \sum_{n=0}^{N} A_n x^n, \quad N \text{ integer} > 0.$$

(i) Obtain a polynomial representation which satisfies the conditions

$y(0) = 0$ $y(1) = 1$
$y'(0) = 0$ and $y'(1) = 0$
$y''(0) = 0$ $y''(1) = 0$

(ii) For what value of x, other than $x = 0$ and 1, is the output acceleration equal to zero?

(iii) Occasionally the designer may impose additional conditions to those in (i), in order to force the cam to produce certain effects.

For example he may insist that in addition to the conditions in (i)

$$y'(0.4) = 0.$$

Obtain the polynomial representation of this new cam.
For what value of x is the jerk zero?

Tutor notes: Part (i) reduces to solution of simultaneous linear algebraic equations.
Part (iii), determination of zero jerk (y'''), requires iterative solution of $f(x) = 0$.

REFERENCES

Cockcroft, W. H. (1982). *Mathematics Counts: Report of the Committee of Enquiry into the Teaching of Mathematics in Schools*. HMSO.
Finniston, M. (1980). *Engineering Our Future: Report*. HMSO.
Lighthill, J. (1979). The mathematical education of engineers, *Bulletin of the IMA*, **15**, 89.
McLone, R. R. (1973). *The Training of Mathematicians*. SSRC Report.

Section C
Use of the Microcomputer and Simulation

16

The Use of Micros in Evaluating and Displaying The Characteristics of Models used in Control Theory

R. V. Aldridge
University of East Anglia, UK

SUMMARY

In classical control theory the bulk of system design is performed in the *s*-plane via Laplace transformation. In general, techniques such as Bode plots and Nichol charts require tedious arithmetic, and the root locus method involves the approximate solution of a polynomial equation. To reduce the tedious aspects of these design methods and hence encourage their proper use, a suite of programs has been developed to run on a BBC microcomputer. This produces graphical output directly on a high resolution colour monitor, and a hard copy on a printer.

Of particular interest is the technique used to evaluate the Root Locus. The method starts by taking the known solutions to be the poles and zeros of the transfer function at the extremes of the parameter range. The intermediate values are then determined by an interpolative solution of the characteristic equation.

1. INTRODUCTION

An area of applicable mathematics that has seen rapid growth recently is control theory. One of the major reasons for this has been the drive towards automation and the growth of robotics in the commercial world.

The first stage in any control theory study is to set down the objectives of the controlled process. It is then necessary to acquire/generate a model that describes the process to be controlled with sufficient accuracy to meet the specifications set out above. The next stage is to set down a controller strategy. Theoretically, there may be many strategies capable of achieving

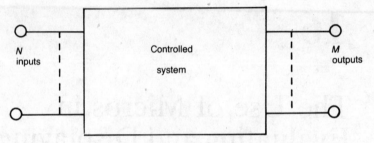

Fig. 1. Multi-input multi-output system.

the desired ends. For obvious reasons the strategy that is normally adopted in industry is the one 'least cost'.

The most common strategy that is adopted in practice is negative feedback. In essence this is because some 'output' measurements can be used to control the 'quality' of future outputs. A typical general multi-input multi-output system is shown in Fig. 1. The model of such a system is normally described by a set of coupled integro-differential (difference) nonlinear equations. The design of the controller for such a system is therefore a very complex process and it is not surprising that it is very rare for an analytic solution to be found. The engineer has to resort to judicious approximations and numerical techniques to obtain workable solutions.

In order that the engineer builds up a good working knowledge his training starts with the simplest of possible systems, that is single input–single output linear deterministic systems. The linearity approximation is not too limiting because, as is well known, any 'stable' system which is perturbed from its equilibrium point can be modelled linearly provided the departures from equilibrium are not too big. However, even after these approximations have been made the derivations of analytic solutions is often tedious and therefore prone to the possibility of error by the young engineer. The purpose of this chapter is to describe the use of microcomputers in aiding the engineer in the first stages of his development. Before going into the description of the techniques developed it is necessary to outline the sort of problems that arise.

2. SINGLE INPUT–SINGLE OUTPUT SYSTEMS

In general this system can be modelled by a linear differential equation with constant coefficients

$$\sum_{j=0}^{n} a_j \frac{d^j y}{dt^j} = F(t) \qquad (1)$$

where y is the output and $F(t)$ some 'driving' term such as an impulse, step, ramp etc. The engineering problem is to determine the appropriate values of $\{a_j\}$ so that the system meets its specification in terms of a given $F(t)$ and initial conditions. Solving (1) is a standard problem in mathematics. The solution consists of a complementary solution to the homogeneous equation, i.e. $F(t) = 0$ plus a particular integral. The determination of the particular integral can be quite cumbersome for non-standard forms of $F(t)$. One technique that considerably reduces the labour is the Laplace transform method. This method is summarised below

$$\mathscr{L}[f(t)] = F(s) = \int_0^\infty f(t)e^{-st}dt \qquad (2a)$$

$$f(t) = \frac{1}{2\pi j}\int_{\sigma-jw}^{\sigma+jw} F(s)e^{st}ds \qquad \text{for } t > 0 \qquad (2b)$$

where $F(s)$ is said to be the Laplace transform of $f(t)$. The great benefit of this method is that (1) is converted to an algebraic equation in s. It is to be noted that in general s is complex. The reason why this method works is that it makes use of the fact that the natural solutions of (1) are exponential in character. The inverse transformation, i.e. back to t is complicated because in general a contour integration is required. However, in practice, enough standard forms have been tabulated to allow the engineer to look up the solution for most situations. Details of how to use the Laplace method can be found in any standard engineering mathematics text (e.g. Kreyszig, 1979).

If $F(t)$ can be expressed in terms of

$$\sum_{k=0}^{m} b_k \frac{d^k f}{dt^k}$$

then (1) can be transformed to

$$Y(s) = G(s) F(s) \qquad (3)$$

where $G(s) = (\sum^m b_k s^k)/(\sum^m a_k s^k)$ is the system transfer function.

The quantity m has to be less than n to ensure physical reality. In the derivation of (3) it has been assumed that all the initial conditions are zero. On substitution of (3) into (2b) it can be seen that $y(t)$ strongly depends on places where

$$\sum^n a_k s^k = 0 \qquad (4)$$

This is a polynomial equation in s. These critical solutions are referred to as the 'poles' of $G(s)$. Equation (4) is in fact the standard auxiliary equation of formal differential equation theory. Physically the solutions of (4) are the natural modes of the problem. In practical testing another feature of $G(s)$ ought to be introduced. This is where $G(s)$ vanishes. This occurs when

$$\sum^m b_k s^k = 0 \qquad (5)$$

These are referred to as the 'zeros' of $G(s)$.

3. FEEDBACK SYSTEMS

The simplest possible feedback system is shown, after transformation in Fig. 2. This is referred to as unity feedback. It can be shown that all feedback systems however complicated can, after manipulation, be

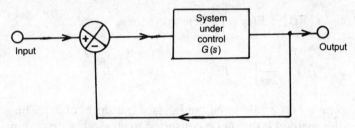

Fig. 2. Single input single output unity negative feedback system.

reduced to this form if $X(s)$ is no longer the actual input to the system but some pre-processed form of it. In this case the effective $G(s)$ of the system can easily be shown to be

$$G_{\text{eff}}(s) = \frac{G(s)}{1 + G(s)} \qquad (6)$$

It is obvious that the poles of G_{eff} are given by

$$1 + G(s) = 0 \qquad (7)$$

and the zeros of $G_{\text{eff}}(s)$ are still those of $G(s)$. In the simplest controllers $G(s)$ is made up of two parts, as shown in Fig. 3, the 'plant' under control and the device used to control it. Normally $G_p(s)$ is fixed, i.e. cannot be

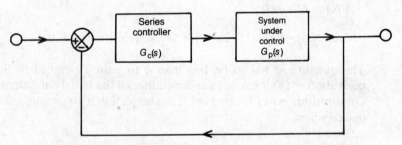

Fig. 3. A simple series controller in a unity negative feedback system.

altered in any way but $G_c(s)$ is at the disposal of the designer. The task therefore is to try to determine a $G_c(s)$ that makes $G_{\text{eff}}(s)$ meet or meet as closely as possible the required design objectives. In classical control engineering $G_c(s)$ is normally of the form

$$\frac{K \prod_{l=0}^{i} (s - Z_l)^{tl}}{\prod_{l=0}^{k} (s - p_l)^{tl}} \qquad (8a)$$

and

$$G_p(s) = e^{-sT} \frac{\prod_{j=0}^{m}(s - Z_j)^{t_j}}{\prod_{j=0}^{n}(s - p_l)^{t_j}} \qquad (8b)$$

where K is called the gain of the controller, s_l is the appropriate pole or zero of order t_l and T the time delay of the plant.

4. TEST PROCEDURES

The specifications of a real control problem are normally covered in terms of the steady state and time dependent properties. The time dependent behaviour is usually specified in terms of the response to certain inputs, for example impulse, step and sinusoidal signals. The model of the plant is usually constructed from measurements made on the system using the above inputs. This latter process is often referred to as plant identification. Once the plant has been identified the controller must be constructed and tested. If this is done in hardware this can often be an expensive and sometimes a dangerous exercise—it is possible for a designer to choose a controller that produces unstable behaviour. This is particularly the case when the plant has inherent time delays. It is better therefore for the designer to test his ideas out on paper before building the test rig. To this end several techniques have been developed to test the designer's conjectures. These are:

(i) The Nyquist plot.
(ii) The Bode plot.
(iii) The Nichol plot.
(iv) The root locus method.

To ease presentation it will be assumed for the rest of this chapter that only linear continuous time systems are being studied. The techniques (i) to (iii) are all related to test signals that are sinusoidal. The methods will only be outlined here. The details can be found in any standard introductory text on control theory (Jacobs, 1974; Richards, 1979; Kuo, 1982).

(i) *The Nyquist plot.* This is a plot of the real and imaginary parts of the $G_c(s)G_p(s)|_{s=jw}$ as a function of w. The shape of this plot gives an idea of the nature of the system and its likely behaviour in real time. Using very general ideas of system stability Nyquist found it was possible to come up with a simple test of system stability. In essence it depends on whether the effective contour on varying w from $-\infty$ to $+\infty$ encloses the point $G(jw) = -1$ or not. If it does then the system is unstable. It is also possible to generate two measures of system behaviour. These are the gain and phase margins. These are illustrated in Fig. 4.

(ii) *Bode plot.* One of the problems associated with the Nyquist plot is that the frequency dependence is implicit. The Bode plot consists of a

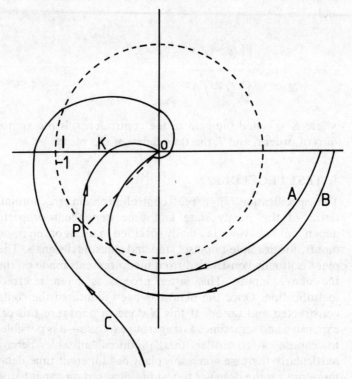

Fig. 4. The Nyquist plot: Curve A is for a stable second order system, B a stable third order system and C for a system with delay and a pole at the origin. C is unstable. For curve B, OK is a measure of the gain margin and the angle IOP is the phase margin.

plot of $20 \log_{10}(\text{amp})$ against $\log_{10} f$. It is possible to set up a stability condition on this plot and to identify the gain and phase margins.

(iii) *Nichol plot*. Once simple stability has been tested with the above plots it is often necessary to determine the closed loop performance from the open loop behaviour. This is the purpose of the Nichol chart for unity feedback systems. The chart consists of closed loop amplitude and phase contours marked off against axis consisting of $20 \log_{10}$ (open loop amplitude) and open loop phase. The closed loop performance is then picked off by looking at the intersections of the open loop trajectory with these contours.

(iv) *Root locus*. As was mentioned in the earlier sections the real time behaviour can be predicted from a knowledge of the poles of a system. For a unity negative feedback system (4), (7), (8a) and (8b) give

$$\left(\prod_{j=0}^{n}(s-p_j)^{t_j}\right)\left(\prod_{l=0}^{k}(s-p'_l)^{t_l}\right) + Ke^{-sT}\left\{\left(\prod_{j=0}^{m}(s-Z_j)^{t_j}\right)\left(\prod_{l=0}^{i}(s-Z_l)^{t_l}\right)\right\} = 0 \tag{9}$$

for the poles of the closed loop.

It is not difficult to see that the solutions are a function of the parameter K. The set of trajectories for fixed p_i, Z_i but varying K is called the root locus. From this diagram it is possible to determine the range of K for stable behaviour and what is more choose the value of K that gives the closest behaviour of the controlled system to the required objective. If these values are not satisfactory the controller poles and zeros can be altered to give better behaviour.

5. MICROCOMPUTERS IN CONTROL

The above process of identification and controller design in the past have depended heavily on one or more of the above techniques. One of the main problems with these methods, especially the root locus technique, was that until the advent of computers they were only very approximate and quite tedious to construct and because of this only relatively simple examples were included in an engineer's training. However, now with the advent of microcomputers with reasonable interactive graphics facilities at low cost these methods can be introduced in a much more accurate and sophisticated form. Because of these features and the speed at which the microcomputers work the student engineers can become acquainted with the behaviour of much more realistic systems before he has finished his undergraduate career. As these methods can be generalised to digital systems the techniques of computer control can also be brought into his armoury.

The speed of the microcomputer enables the engineer to examine several design variations for his controller before resorting to the hardware prototype. It enables the root locus method to become a quantitative design tool.

6. THE 'CONTROL AIDS' PACKAGE

The rest of the chapter will concentrate on a particular suite of programs mounted on a BBC model B microcomputer using a C6502 second processor. The latter is not necessary for the suite but was used to speed up the calculations. All the graphical output was produced in the high resolution four colour mode 1. Screen dump facilities were available to an EPSON MX80 printer.

The original aim of the package was to produce software that supported the second year course in control theory given to electronic systems engineers. The general idea was that the student supplied the information concerning the system in terms of a model in the Laplace domain in the form of the position and nature of the poles and zeros and any associated time delays. The software was then constructed so that graphical output could be produced for any of the above plots plus the ramp inputs. Actually as the package was developed it was realised that not only could any plots be demonstrated but that proper interactive design could take place. The suite structure is shown in Figure 5. The algorithms for constructing the

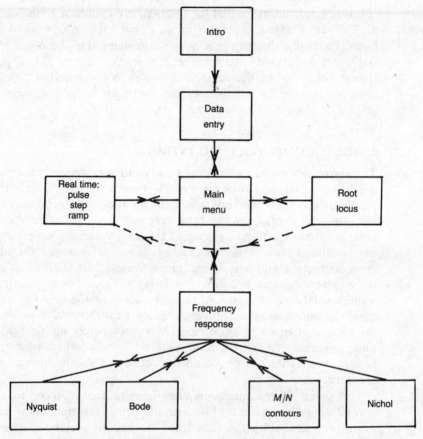

Fig. 5. The CONTROL AIDS program suite.

Nyquist, Bode and Nichol plots is fairly straightforward. Essentially they consist of substituting in different frequency values via the relationship $s = jw$ and then working out the various plotting values for the chosen plot Examples of Nyquist and Bode plots are shown in Figs 6 and 7.

The root locus plot is more interesting. In principle (9) is a very difficult problem to solve because it is a transcendental equation. However, an examination of (9) quickly shows that when $k \to 0$ the solution must be the poles of the open loop and that when $k \to \infty$ the solutions must either be given by $e^{-sT} = 0$ or the zeros. It also can be seen that as k is varied from these extremes that the number of branches leaving a singularity is equal to the order of the singularity. Once this has been realised it is straightforward to see how the rest of the root locus can be constructed. The method is based on the fact that (9) can be rewritten in terms of the magnitude and phase of each complex vector. The solution of (9) then becomes a condition on the magnitude and phase of the solution values.

There are many ways this can be solved numerically. The two that have been tried with equal success are Newton Raphson and a simple 'rotating' vector approach. Typical output is illustrated in Figs 8 and 9.

The Use of Micros in Control Theory   213

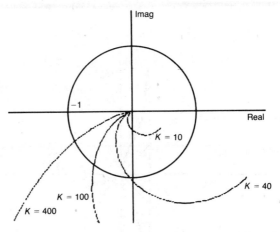

Fig. 6. The Nyquist plot for two simple poles at $(-3,0)$ and $(-6,0)$ on the S-plane for various values of K.

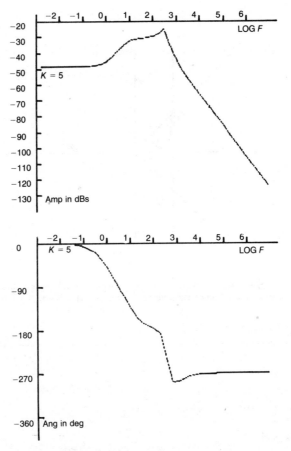

Fig. 7. The Bode plots for a system with singularities at: poles $(-8,0)$; $(-100, 300)$ and $(-100, -300)$ and zeros $(1,0)$ and $(-600, 0)$.

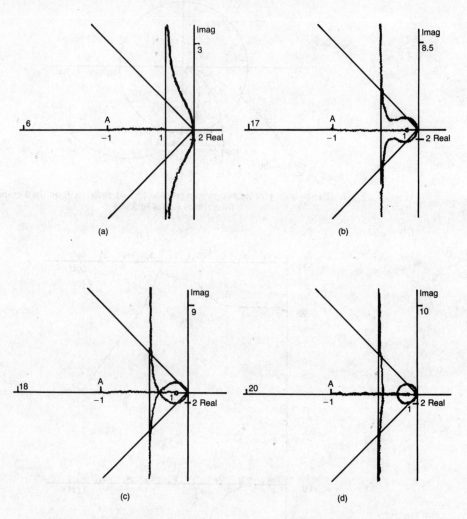

Fig. 8. The root locus for a system with a double pole at the origin, a zero at $(-1, 0)$ as a function of a simple position at A: (a) A is $(-3, 0)$; (b) $(-7.5, 0)$; (c) $(-8, 0)$ and (d) $(-10, 0)$.

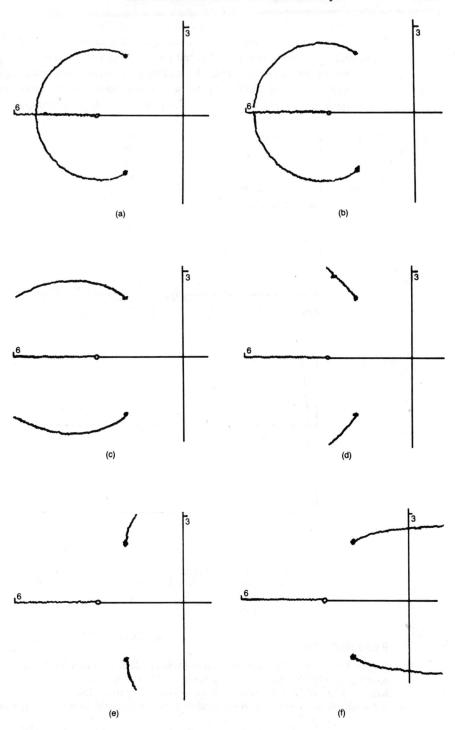

Fig. 9. The root locus variation for a second order system with two poles at $(-2, 2)$ and $(-2, -2)$ and a zero at $(-3, 0)$ with time delay. The time delays are: (a) 0; (b) 0.05; (c) 0.1; (d) 0.3; (e) 0.5 and (f) 1.

The suite also has the facility for working out the real time response for both open and closed loops of systems for the standard test inputs. The pulse and step response are illustrated for a second order system in Fig. 10.

The use of microcomputers in the teaching of control theory to engineers has greatly enhanced their appreciation of the power of the basic tools of the trade and enabled them to use them in an intelligent manner.

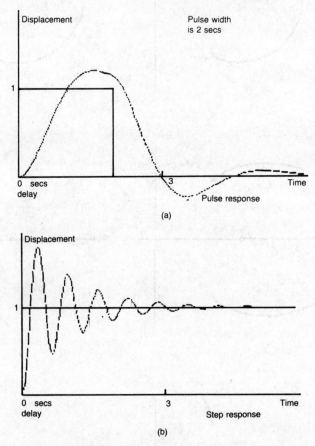

Fig. 10. The pulse response for a second order system with poles at $(-1, 2)$ and $(-1, -2)$ and the step response for a system with poles at $(-1, 10)$ and $(-1, -10)$.

REFERENCES

Jacobs, O. L. R. (1974.) *Introduction to Control Theory*, Clarendon Press, Oxford.
Kreysig, E, (1979). *Advanced Engineering Mathematics*, Wiley.
Kuo, B. C. (1982). *Automatic Control Systems*, Prentice-Hall.
Richards, R. J. (1979). *An Introduction to Dynamics and Control*, Longmans.

17

Mechanics with a Micro.

C. E. Beevers
Heriot-Watt University, Edinburgh, UK

SUMMARY

The teaching of Newtonian mechanics is limited to a relatively few Scottish schools and the consequence is that many undergraduates meet mechanics for the first time as a mathematical discipline when they arrive at university. Few of the students are confident at the start with the new subject and so any fresh ideas to aid motivation are welcome. I have found that the microcomputer is an ideal teaching aid for Newtonian mechanics. Over the past two years I have developed suites of software to emphasise and expand on the material already presented in a more conventional way. The programs fall into three categories: simulation, 'games' and teaching. The students find that simulations of real physical problems are most useful in helping their understanding of the underlying principles. Some of the concepts of mechanics can be illustrated within a simple 'game'. Other topics have to be covered by a straightforward teaching approach but even in such programs the computer has a unique role to play. Mechanics has been an area of traditional difficulty for Scottish students of mathematics but I have found that their interest and ability to cope with the problems of the subject have been enhanced by the use of the computer. In this chapter I describe some of the features of the software and some of the reasons for producing it.

1. INTRODUCTION

Newtonian mechanics is the mathematical model proposed by Newton to study the motion, and the causes of motion, of rigid particles. It is a subject that has traditionally caused difficulties to students of Mathematics. The principles of mechanics can be taught in an introductory course which considers the basic problems of motion for a single rigid particle. Even then Scottish students find the mathematical discipline of mechanics a hard one with which to cope. For, in many Scottish schools, mechanics is a subject encountered only in Physics and when students meet it for the first time as a mathematical subject they are frightened by it and find it difficult. Any means to combat or reduce this fear and encourage the students to become independent learners are to be welcomed.

I have taught a course of Newtonian mechanics to first year undergraduates at the Heriot-Watt University in Edinburgh for ten years now. Some two years ago I started to introduce some simple computer programs into the course to consolidate and enhance the more conventional approach to lectures. This software was also designed to allow the students to study the subject at their own pace. I have written software which falls into three main categories as follows: simulation software, 'games' programs and teaching packages. These three different software strategies provide three distinct responses to the problem of teaching Newtonian mechanics. In this chapter the software design of the programs in each category is discussed.

In section 2 some of the simulations are described and the student comment is analysed. Plato says in his book *The Republic* that 'No compulsory learning stays in the soul. In teaching children train them by a kind of game and you will see the natural bent of each'. This old philosophy is one which is accepted by a number of authors including among others O'Shea and Self (1983). It is also acknowledged by no less a body than the Computer Board of the UK and in their report in December 1983 they point out the role of a 'game' in the learning process. In section 3 some of the details are explained of the software games which illustrate the principles of mechanics.

Section 4 looks at what I consider to be straightforward teaching packages. There are some subjects that do not lend themselves to a more exciting approach but I am convinced that there is still a place for such programs to back up the conventional lecture material. In all three types of software the computer has a unique role to play.

The final section seeks to draw the threads together by emphasising the main aims of this approach. This work is intended to provide an alternative form of learning experience, to achieve a deeper understanding and revision of mechanics and to encourage the independent learner. The software enables the student to work on his or her own in a more stimulating and self-checking way and it seeks to promote the confidence of the user in handling the principles of mechanics.

2. SIMULATION SOFTWARE

The computer has the power to provide a graphic description of a simple experiment in a subject like mechanics where such experiments would be difficult or expensive to set up. This allows the student to 'see' what the mathematical equations are predicting. Moreover, it gives the student the chance to try out different inputs in the form of initial conditions to discover how this affects the outcome of the particular experiment.

One of the simulation programs involves a simple pulley problem. Two particles of masses M and m are on either side of a fixed pulley and are attached to each other by a light inextensible string which passes over the smooth pulley. The system is released from rest and the larger particle descends towards an inelastic plane which it strikes and comes immediately to rest. The smaller mass continues upwards under gravity and it, too, stops before falling back to jerk the larger mass off the plane. The jerking motion indicates an impulsive force and both particles then move together again. This problem has three distinct phases with each phase governed by a different set of equations. There is conservation of energy and constant accelerated motion in two of the phases and energy loss but momentum conservation in the third phase. So, although this problem is not very exciting in itself the principles of mechanics are well represented within it. The simulation enables the student to see these different stages and to look at them over and over again if necessary. Once the program has been set up it will repeat its message to the student again and again and unlike the patience of the human teacher this artificial tutor continues to explain the distinct parts of the motion as long as the student requests it. So the computer has two advantages over the human teacher: its patience to repeat the arguments and the visual display it offers. In addition, the program can be constructed to give the students the opportunity to input their own values for M and m and this, too, is instructive and valuable in the learning process.

There are other situations that lend themselves to simulations and enable students to improve on their understanding of the underlying principles of mechanics. For example, in circular motion the problem in which a particle is initially disturbed from the highest point on the smooth outer surface of a sphere can be simulated. The prediction of where the particle loses contact with the sphere can be graphically illustrated. This problem is ideal for graphic simulation and a program can be written to describe the subsequent motion of the particle after it has left the sphere.

Motion in a resisting medium and the speed of a ring attached to an elastic string in which the ring is constrained to move on a vertical wire are further simulation possibilities. In each of these situations the user is able to input some initial conditions and see how that influences the outcome. Moreover, in such examples the computer can be set up to provide some dynamic programming which aids the student's solution of a specific problem. It is the ability of the computer to bring the subject of mechanics to life that appeals to the student. All of them agree that it is the visual display that is the most important feature of these programs. I have adopted the view in designing the software that each program should be complete in itself and not require a separate printed explanation. The instructions for use should be clear and appear on the screen as the program progresses so that it is obvious what to do next. A questionnaire given out at the end of the course confirms that the students do find the programs easy to use.

3. SOME SIMPLE GAMES' PROGRAMS

I was not aware of Plato's assertion when I started to produce some simple games to illustrate different aspects of mechanics. But it is clear that by gripping the students' attention with a simple game it intrigues them until they have found out how it works. By this time there is some mechanics in the 'soul' and later in the year it may even appear in an examination script!

As a subject mechanics is ideal for the construction of games programs and most of the examples in this section are taken from the book by Burghes & Downs (1975). For example, in designing a two-stage rocket an optimal final speed is achieved by solving a quadratic equation with the ratio of the mass of stage 2 to the sum of the masses of stages 1 and 2 as the unknown. This can be set up as a game in which the student has to bring the rocket from Mars to Earth and this can only be done if the ratio is optimal in the sense described. A graphical display in this problem and its interactive possibilities encourage the students to tackle a problem they might otherwise avoid.

There are other situations that can be used to illuminate the principles of mechanics and the next three examples are taken from the article by Beevers (to appear). Orbital motion is rich in interesting examples. The principles of angular momentum and energy conservation form the basis of the equations governing orbital motion. These principles can be well illustrated within a simple game. Take the problem of a satellite S on a circular orbit above the surface of a planet P. By reducing the speed of S as it circles P the path of S becomes elliptic taking S closer to P. If the correct choice is made and the closest approach is equal to the radius of the planet then a safe landing is assured. Such a program has much to say not only on the geometry of the ellipse but also on the principles of Newtonian mechanics.

The motion of a ballistic missile in modern warfare provides another possible 'game'. For, the range of such a missile is a function of the angle of projection called the heading angle. So, to direct the missile on its optimal trajectory the student must differentiate an expression for the range in terms of this heading angle. This game has scope for drama and is topically called GREENAM (Greenham is too long to be recognised as a program name on the BBC micro).

The problem faced by a fireman trying to direct water through a broken window at a height H above him into a burning building forms the basis of another game. The fireman must stand as far from the foot of the blazing inferno since the heat is so intense and the wall may collapse. The object of the game is to find the angle of projection which maximises the horizontal range. The motion of projectiles in general is full of examples which with a little ingenuity can be turned into simple games.

In all these cases the computer enhances the problems by enabling the parameters in the problem to be varied by a randomly generated sequence. In each game it is only when the mathematics is properly understood that the student is able to 'win' the game every time. If the student does not understand a particular game then there are further programs to explain the mathematics of the games. As before it is important to produce software that the student finds easy to handle so that they are confident to try out the material.

4. TEACHING PACKAGES

The third category of software is contained in a straightforward teaching package in which a particular topic is explained in a manner reminiscent of a lecture. However, such material on the computer has a number of distinct advantages. For example, it can be covered at the student's own pace with the program progression under the control of the user. Further, worked examples can be laid out and the main points in a method highlighted by means of the computer facilities of sound or colour. This teaching technique leaves an imprint in the mind of the student at the important steps through a method. It is within the subjects like differential equations or vector algebra where methods are relatively straightforward that this type of material is most useful. It also has a place, though, in underlining the main lines of development through a question on inverse square law orbital motion. Humour, too, can be built into each program to emphasise the important points.

There is, again, the computer feature of being able to generate a random number. This allows each teaching package to conclude with some questions to the student which are of a type but different in detail. This helps to preserve the freshness of each package when exactly the same question does not appear until the program has been used a good number of times. Finding particular integrals in a second order ODE with constant coefficients is an ideal example of the type of subject that can be best handled by means of a teaching package.

5. CONCLUDING REMARKS

The computer is an ideal teaching aid. It will never replace the human teacher but it does provide an excellent back-up to conventional lectures and tutorials particularly in courses of mathematical modelling. So, the micro is having a growing influence in a subject like Newtonian mechanics where graphic display plays such an important role in a proper understanding of the material.

The software strategy in a particular case depends on the nature of the topic to be explained. Simulations, games and teaching packages all have their parts to play. With sound and colour facilities the computer can illuminate the main steps in a method. Through games and simulations most subjects can be brought to life and the micro now becomes an essential part of the teacher's tool-kit.

Finally, the computer's ability for great variety is a feature that ensures the freshness of each package in which the same problem recurs infrequently. The programs enhance the teaching of mechanics in a way that encourages the independent learner. The features of self-checking and interactive response provide an aid that is hard to equal. In all these ways, then, the programs provide an alternative means of learning and help to deepen the students' understanding of mechanics. All the programs described in this chapter will be appearing in the text by Beevers (in preparation) though details of the software can be obtained directly from the author on request.

REFERENCES

Beevers, C. E. Motivating mechanics, *IMA J* on the Teaching of Math. and its Applications (to appear).
Beevers, C. E. BASIC Particle Dynamics (in preparation).
Burghes, D. N. & Downs, A. M. (1975). *A Modern Introduction to Classical Mechanics and Control*, John Wiley, London.
O'Shea, T. & Self, J. (1983). *Learning and Teaching Computers*, Harvester Press Ltd, John Spiers, Brighton.
Report from the Computer Board of the UK, December 1983.

18

Computer Applications of Modelling for Mechanical Engineers

G. E. Beswick

and

A. S. White

Middlesex Polytechnic, UK

SUMMARY

The academic year 1984–85 saw the introduction of a course in computer applications in our mechanical B.Eng course. This course is of novel conception in that it involves engineering and mathematics lecturers teaching the course both in parallel and tandem.

Areas of work covered by the course include microprocessors and microcomputers, use of library subroutines such as NAG, CAD, Numerical Analysis and the modelling packages PAFEC and ACSL. The lectures on numerical analysis are run in tandem while all the others are run in small groups in parallel to minimise the hardware load on our DEC10 and PRIME computers.

This chapter outlines the rationale behind the modelling part of the course with PACEF and ACSL and illustrates the achievements of students and how the programs are tied together to compare their relative merits. Since the major role of the course is to discuss applications several pertinent examples are given. Assessment of the students is also discussed.

1. INTRODUCTION

All CNAA engineering degrees are undergoing radical transformation in the light of the Finniston report (1980) and the pressure of new technology. The old BSc format is being replaced by the new B.Eng schemes. These contain a far greater proportion of applications than before. This should not just be seen as 'bolt on' extras but as a radical transformation of the whole approach to engineering education. The B.Eng schemes, as HMI (1985) maintain, are to be truly computer integrated engineering schemes with CAE permeating the whole course in every subject. Our own approach to this procedure is to insert into the first year of the course the necessary introduction to computer programming and then, in the second year, as the major plank of the integration process to include a subject entitled 'Computer Applications of Engineering'. This is not, as its title might suggest, the only area where the student is intended to encounter CAE. He is expected to meet it also in the relevant skills which he will apply to other subjects. It was, however, considered to be more economical to centralise the instruction of various techniques and introduce subject expertise in one course rather than the other way around. Thus five areas of work were grouped together in this course (Fig. 1) namely Computer-aided Draughting (CAD) using the MEDUSA system, a microprocessor applications module, use of NAG subroutines, solution of Stress and Heat transfer problems using the PAFEC Finite Element package and the solution of time-dependent problems using ACSL which is a continuous simulation package. The work described in this chapter is concerned with the last two elements, that is the modelling part of the course.

Most engineering courses spend much time on experimental modelling or iconic simulation. Many courses used to spend time covering analogue simulation. We have replaced the analogue approach completely by digital simulation.

To quote Rosko (1964):

This course is concerned with the mathematical simulation in which the Engineer utilises a mathematical model (an abstract system) to represent

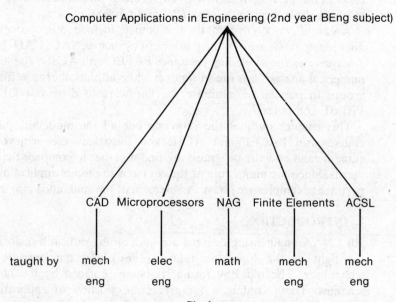

Fig. 1

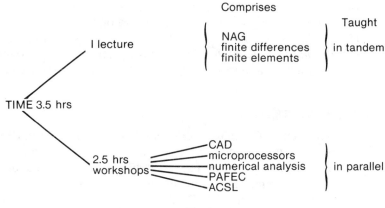

Fig. 2

a physical system. Through this expedient he is able to obtain and evaluate information concerning the system's responses to various excitations or stimuli and to predict its behaviour under a given set of conditions. Mathematical simulation has as its objective, as does iconic simulation, the economic representation of complex physical systems, so that they may be studied easily and conveniently. Simulation today is often a required step in current sophisticated research and development programmes for it is usually the only way in which engineers may analyse different versions of a complex physical system at a fraction of the system's aggregate cost.

2. OBJECTIVES

The primary objective of this course is to enable students to become aware of the various techniques of computer usage in engineering and to be able to use them well enough to solve medium difficult problems in their final year. For the proposed final year of the B.Eng course with which we are involved this design exercise could probably surface in the student's individual project and almost certainly in his group design project which is to be one of the more salient elements of his final year.

Subsidiary objects of the course are to enable students to utilise the packages for laboratory and design work in their second year.

Abilities to be tested

B. S. Bloom (1964) lists the abilities required by the student.

(1) *Knowledge*—ability to recall facts and practical techniques
(2) *Comprehension*—ability to translate data from one form to another, in our case to convert the physical problem into a computer model.
(3) *Application*—the ability to apply knowledge, experience and skill to new situations, in our case to use modelling techniques on a range of problems.

(4) *Analysis*—the ability to break down the elements of presented material, recognise their relations and organisation.
(5) *Synthesis*—the ability to put the parts together to form a whole.
(6) *Evaluation*—the ability to make a judgement as to the value of the information in terms of either internal or external criteria.

3. FACILITIES

The Polytechnic Computer Services run at present a DEC10 mainframe computer with 130 connected terminals plus a graphical facility for PAFEC output only. They also have a PRIME mainframe for driving the MEDUSA system. The core restraints implicit in the DEC10 mean that the full power of both PAFEC and ACSL cannot be used. The Polytechnic will, however, be receiving 2 IBM 4381 computers, 2 VAX computers and one Data General computer. It is at present planned to run both PAFEC and ACSL on the IBM set up by December 1985. This should enable the full power of both packages to be utilised on line.

4. TEACHING METHOD

Because of the great differences in the type of work being covered the approach in the individual areas is different but the course is split into two parts—lectures and workshops. The weekly lectures are given to the whole class while the workshops operate in groups of 8 students with five sessions for each group for each area. This was done for two reasons. The first was to limit the call on computer resources. The second was to enable efficient communication to take place while demonstrating the packages. The duration of the workshop is two and a half hours each week for each group. (see Fig. 2).

In the lectures basic topics of numerical analysis are covered plus an introduction to the techniques of finite differences, finite elements and boundary elements. For the last two techniques the application is limited to potential problems. This is because of the relative simplicity of the mthod in this area and also the large class of physical problems, drawn from all branches of engineering, which reduce themselves mathematically to the solution of the potential problem with simple boundary values.

In the workshops both ACSL and PAFEC are taught as 'black-box' routines. Graded problems are tackled by each student thus introducing him logically to the various segments of the package he will need to understand to be able to use them economically in his second and, more importantly, his third year on the course. At the same time the student is encouraged to analyse both the input—is it sensible? i.e. is the problem he wishes to solve couched in the appropriate form—and the output. The latter is stressed as it is a common habit for most students to produce a mass of output data which is not examined critically. Phrased succinctly the student is encouraged to be confident using the packages while at the same time being forced to realise that the methods are approximate and only work as well as the approximations adhere to the reality of the problem.

4.1. Computing point of view
We choose at this point to consider PAFEC and ACSL separately.

First Domain. Familiarity with the package syntax. We use a simple example for which we have an answer from other methods.

PAFEC
Simple plate with a hole under tension (Fig. 3).

Choice of elements and nodes to allow easy mesh refinement and choice of graphical output extension to bending problems and those of thermal stresses (Fig. 4).

Point and pressure loads.

Use of 3D elements for the problem of a solid sphere (Fig. 5).

Use of more complex Pafblock specification for element generation plus use of restraints module to allow use of symmetry.

ACSL
Simple pendulum.

How to vary the integration. Step size to produce stability of solution, output facilities, graphs etc., ease of inserting non-linear terms (Fig. 6).

Aerodynamic oscillation problem.

Non-linear impact and divergence, with state event function extension into other fields by use of control system simulation.

Second Domain.

PAFEC
Taking a design problem and choosing a suitable set of nodes/elements for stress evaluation with a time limit and to analyse the resultant output. Emphasis is placed on using an efficient method of solution (Fig. 7).

ACSL
Obtain an optimum solution. Students are given an engine vibration problem and asked to predict the worst engine speed and design a vibration absorber to eliminate the problem. No time limit set (Fig. 8).

4.2 Engineering point of view
The basic problem of modelling for mechanical engineers is in understanding the limits of the assumptions behind the process and evaluating the solutions generated by the PAFEC. and ACSL packages. The advent of the use of digital computer-based packages like the ones we are describing has wrought a radical departure from the older methods of problem analysis. Instead of having to analyse small packets of data the user is now almost swamped with output. We therefore place great

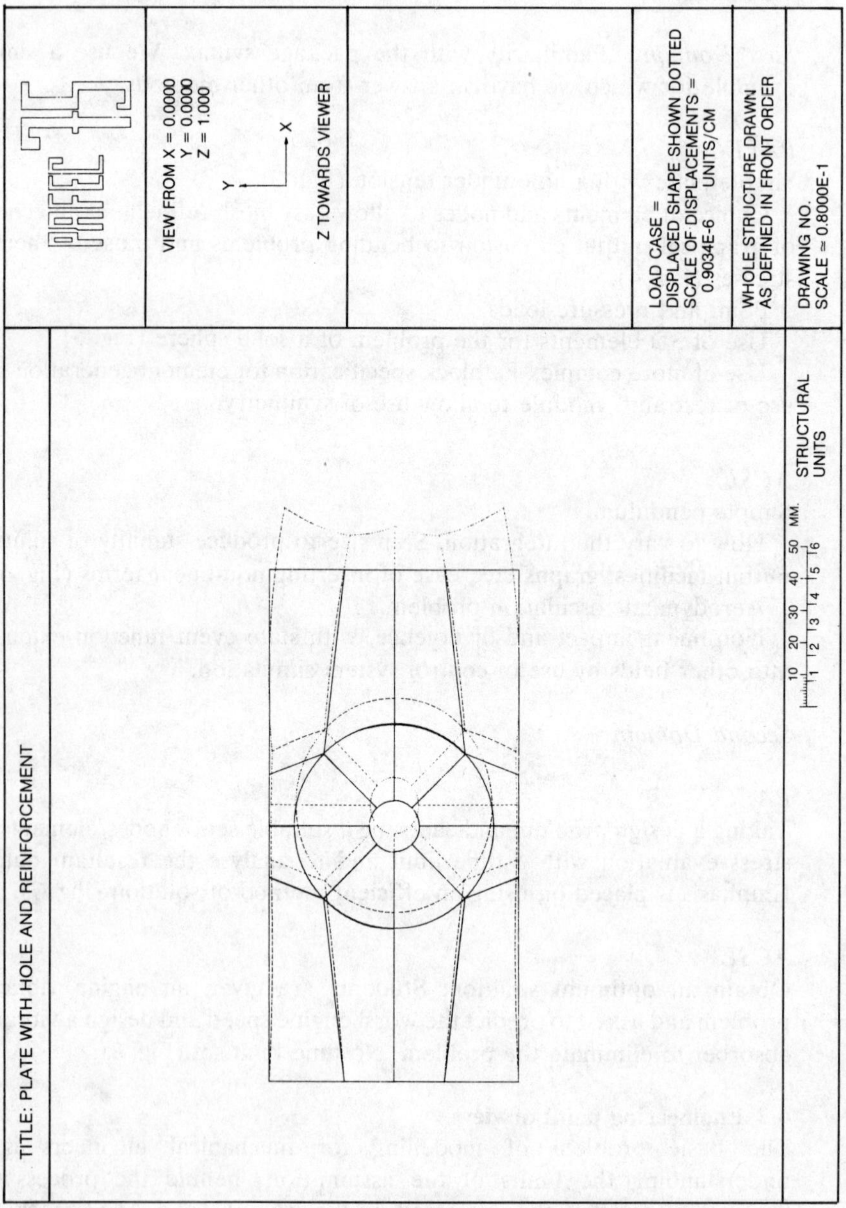

Fig. 3

Computer Applications of Modelling for Mechanical Engineers 229

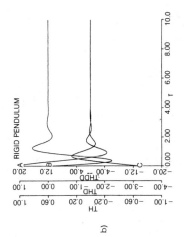
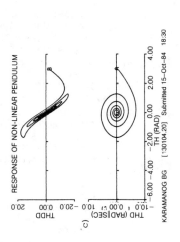

Fig. 4 – Heat exchanger simulation.

```
.TY ACSL3.CSL
PROGRAM HEAT EXCHANGER
        "SIMULATION OF A HEAT EXCHANGER WITH MEASUREMENT LAG AND"
        "VALVE DYNAMICS"
        CONSTANT KC=1.0,KV=1.0,S=1.0
        CONSTANT KE=1.0,TV=3.0,TSTP=50.0
        CONSTANT TDE=6.0,T1=17.0,THIC=0.0
        CONSTANT T2=6.0,TDM=1.0,VOIC=0.0
        CONSTANT TM=12.0
INITIAL
        P=T1+T2
        Q=T1*T2
END$ "OF INITIAL"
DYNAMIC
        CINTERVAL CINT=0.1
        NSTEPS    NSTP=1
        MAXTERVAL MAXT=0.1
        "SELECT COMMUNICATION INTERVAL AND STEP LENGTH"
DERIVATIVE
        "PLANT MODEL"
        CO=KC*E
        VO=KV*REALPL(TV,CO,0.0)
        TH=CMPXPL(P,Q,VOD*KE,0.0,0.0)
        THM=REALPL(TM,THMDEL,0.0)
        "INCLUDE TIME DELAYS IN EXCHANGER AND MEASUREMENT"
        VOD=DELAY(VO,VOIC,TDE,100)
        THMDEL=DELAY(TH,THIC,TDM,100)
        "INPUT A STEP"

        THC=S*STEP(0.1)
        E=THC-THM
END$ "OF DERIVATIVE"
TERMT(T.GE.TSTP)
END$ "OF DYNAMIC"
END$ "OF PROGRAM"

.TY FOR20.DAT
1ACSL RUN-TIME EXEC   VERSION 5 LEVEL 8F   19-Mar-85          18:13    PAGE   1


        PREPAR T,TH,THC
        SET TITLE="HEAT EXCHANGER"
        SET CALPLT=.T.,TSTP=250
        PROCED HEAT
        SET DEFPLT=.T.,KC=5.0
        START
        PLOT "XAXIS"=T,"XHI"=TSTP
        START
        PLOT TH,"HI"=1.5,"LO"=0.0,THC,"SAME"
        SET DEFPLT=.F.,KC=3.0
        START
        PLOT TH,"HI"=1.5,"LO"=0.0,THC,"SAME"
        END
        HEAT
        SET DEFPLT=.T.,KC=5.0$
        START$
        PLOT "XAXIS"=T,"XHI"=TSTP$
        START$
        PLOT TH,"HI"=1.5,"LO"=0.0,THC,"SAME"$
1ACSL RUN-TIME EXEC   VERSION 5 LEVEL 8F   19-Mar-85          18:13    PAGE   2
        HEAT EXCHANGER
```

Fig. 5

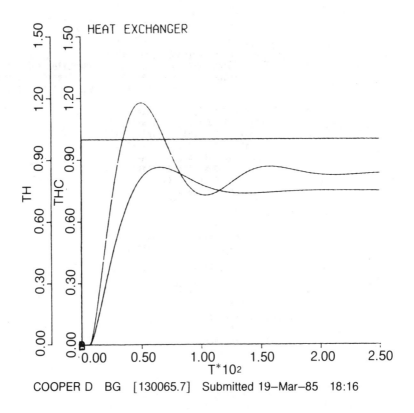

Fig. 5 (continued).

emphasis on graphical output and histograms as a method of identifying crucial areas and combinations of loads, speeds, etc. Previous engineering courses put the emphasis on the most difficult aspect of design, i.e. devising a method for producing performance output. The arrival of packages on the educational scene is to put the emphasis where it most matters—using the data to make *design* decisions.

4.3 Evaluation of performance

Although the course is labelled as being continuously assessed it is not totally assessed in this manner, neither is it really a set of objective tests. Finniston calls for the accentuation of making 'real' design decisions—the student's goal is to satisfy course requirements that will enable him/her to proceed to the final year of the honours course. In our view too much formal assessment will lead him away from the understanding of what engineering ought to be about and into avenues whose sole outcome is arid mark-accretion.

We really want them to be able to run the packages to obtain solutions to problems. While we would like them to understand all the output and the

```
TY TEST.CSL
PROGRAM INTERNAL COMBUSTION ENGINE
        "SIMULATION OF AN INTERNAL COMBUSTION ENGINE"
        "MOUNTED ON ATEST PLATFORM WITH VERTICAL DISTURBING"
        "FORCE CAUSING VERTICAL DISPLACEMENT OF THE TEST BED"
INITIAL
   CONSTANT M1=220.5     ,M2=14.5
   CONSTANT D=1.83       , A=0.075
   CONSTANT EI=1.24E6    ,PI=3.142
   CONSTANT YDIC=0.0     ,YIC=0.0
   CONSTANT TSTP=5.0     , N=1500
        K=(M2*A*(PI**2))/(900.0*M1)
        B=(48.0*EI)/((D**3)*M1)
        F=(PI)/(30.0)
END$"OF INITIAL"
DYNAMIC
        CINTERVAL       CINT=0.005
        MAXTERVAL       MAXT=0.001
        NSTEPS          NSTP=1
DERIVATIVE
        YDD=((K*N*N)*SIN(F*N*T))-(B*Y)
         YD=INTEG(YDD,YDIC)
          Y=INTEG(YD,YIC)
END$"OF DERIVATIVE"
END$"OF DYNAMIC"
TERMT (T.GE.TSTP)
END$"OF PROGRAM"



TY TEST.CTL
.R ACSL
*TEST
.DO ACSLNK TEST.CAP
*S CALPLT=.T.
*S GRDCPL=.T.
*PREPAR T,Y,YD,YDD
*START
*S N=1600
*S TITLE="ENGINE TEST BED;N=1600"
*PLOT Y
*START
*S N=1700
*S TITLE="ENGINE TEST BED;N=1700"
*PLOT Y
*START
*S N=1800
*S TITLE="ENGINE TEST BED;N=1800"
*PLOT Y
*START
*S N=1900
*S TITLE="ENGINE TEST BED;N=1900"
*PLOT Y
*START
```

Fig. 6. Pendulum problem.

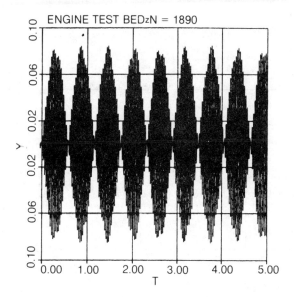

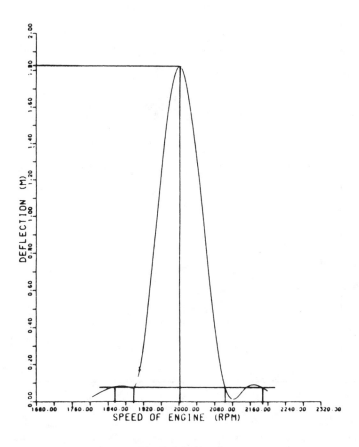

Fig. 6 (continued).

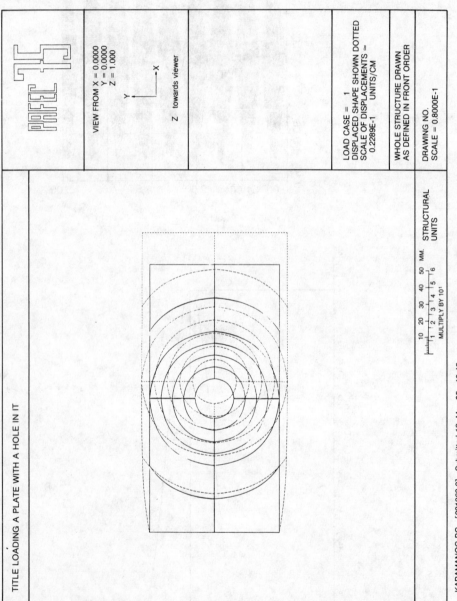

Fig. 7.

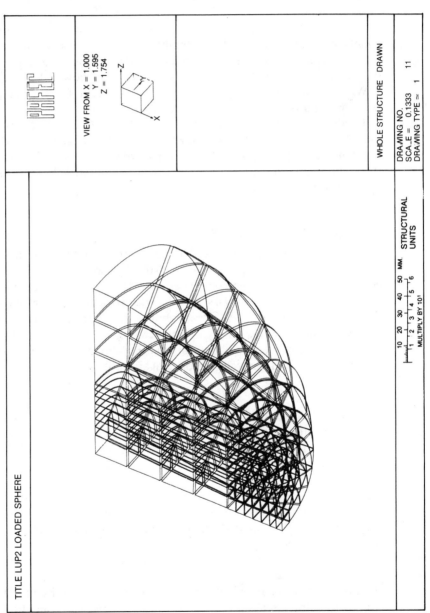

Fig. 8. Engine vibration.

engineering assumptions on which it is based we are not sure this is possible within the time constraints placed upon this segment of their second year. Sections 1, 2 and 3 can be achieved. Sections 4 and 5 have been partly achieved at the time of writing and 6 has only been achieved by a small number of students.

In the PAFEC package we have only touched upon a small fragment of what the package is capable of. Fundamental to all aspects of this tool, however, is the ability to select *ab initio* a node/mesh arrangement that can be refined easily about areas which we intuitively know to be critical.

With ACSL the use of the package is different for solving control problems and regular differential equations.

To allow for this difference between the PAFEC and ACSL packages only one formal exercise was set for the PAFEC package while five ACSL problems have been set with increasing order of complexity. The final ACSL problem is a real design exercise and has therefore been given more weight (40%).

5. PROBLEMS WITH RUNNING THE COURSE

Generally we found that the overall demand for computer access has grown far beyond what the DEC10 can realistically handle. Packages demand large amounts of core and sometimes have to be run 'on-line' and not in batch mode. The relative small size of the DEC10 has been a constant source of frustration to both students running the packages and to staff demonstrating them. Hopefully with the coming on stream of the two IBM computers this bottleneck should be considerably eased. As to problems associated with each package we once again split them into two

PAFEC

Students had difficulty in choosing correct mesh generation. A usual fault, upon mesh refinement is to generate elements which are too thin or have too small or too large corner angles causing the program to fail. Core restrictions mean that only approximately 130 3D elements can be generated.

Shear stress output for 3D elements not available. Also the application of pressure loadings is very awkward and for some 3D elements impossible at this level (level. 2) of PAFEC. Hopefully we shall soon receive level.5 version of PAFEC.

ACSL

Despite being warned students chose an integration step size which is too large (surprisingly seldom too small) causing numerical instability. Some students put in physically unrealistic data (Fig. 6b).

The students are not able to run ACSL on line until after five o'clock each day. A lot of time is absorbed in running in batch mode overnight. As a new package on the system there have been far too many operating 'bugs' as we all have found to our cost.

6. COMMENTS AND CONCLUSIONS

The main area in which we see great problems for the future is the tendency of the students to become too package-reliant and with a lack of ability to evaluate the implications of the output data. When they come to use these packages in their several design modules of the final year of their degree this tendency should be corrected. One way to counter this is to design sophisticated laboratory experiments where the only way to tackle the problem is via the use of the two packages and to check the output from them against experimental data.

As was to be expected there was great variation amongst the students in their ability to interface with the machine and the attendant packages. Some students soon acquired a 'feel' for how the package worked whilst others seemed altogether lacking in understanding of the basic operations of a digital computer. Most, however, became reasonably proficient at putting problems on the machine.

Finally we would like to warn other teachers attempting a similar exercise to make sure they have a big enough and fast enough machine with good graphical output for interactive access.

REFERENCES

Bloom, B. S. (1964). *Taxonomy of Educational Objectives*. Longman.
Finniston, Sir M. (1980). Report of the Committee of Enquiry into the Engineering Profession, HMSO.
HMI (1985). An HMI Perception of the Impact of New Technologies on the Practice of Engineering, Polytechnic of Central London 1–3 April 1985.
Rosko, J. S. (1964). *Digital Simulation of Physical Systems*. Addison-Wesley.

19

Mathematical Modelling using Dynamic Simulation

R. R. Clements
University of Bristol, UK

SUMMARY

One important aspect of the practical use of mathematics in the solution of 'real world' problems is the choice of a mathematical formulation which is appropriate to the problem. Ideally the mathematical model of the real problem should be sufficiently complex and detailed to yield results that are of importance in the solution of the real problem (as opposed to the solution of the mathematical model of the problem) yet sufficiently simple that results are not obscured by unnecessary detail and the mathematical solution process is neither too tedious nor too expensive.

The development of the skill of choosing a level of modelling that is appropriate to the problem has been characterised by the author in previous papers as the development of 'mathematical discretion'. It has been noted that inexperienced modellers usually choose too detailed a level of representation rather than too superficial a level. This is often caused by insufficient attention to developing a physical understanding of the problem before tackling its mathematical formulation.

Simulation systems can play a vital role in developing students' powers of mathematical discretion. They offer a simple route to studying a range of possible models, of varying degrees of complexity, for a physical problem and comparing the results. A well structured course involving the study, by simulation, of a variety of physical and organisational problems can be very valuable in developing, in students, the ability to choose appropriate mathematical representations of problems. The fund of experience, albeit vicarious, which is thus built up improves students' performance in subsequent modelling exercises.

This chapter illustrates, via a study of student response to a representative exercise in modelling a physical system, a typical range of mathematical formulations that may be offered and comments on their usefulness in solving the real problem. A dynamic simulation system for the BBC Microcomputer is briefly described. The simulation system, BCSSP, will be demonstrated in conjunction with the chapter.

1. INTRODUCTION

The simulation/case study technique for facilitating student learning in the area of the application of mathematics to the solution of problems arising in the 'real world' has been developed by the author over a period of nearly ten years and actively used with Engineering Mathematics degree students at Bristol for seven. When the course was first conceived objectives for the course were set and are reported in Clements & Clements (1978) and in Clements (1978). Two of these objectives were

(a) To give students practice in evaluating the effects of various sections of, and inputs to, models, and making appropriate simplifications and approximations to aid efficient solution
(b) To give students practice in critically examining the various possible approaches to, and models of, a system, and choosing an optimal or near-optimal method of analysis within the constraints of the system.

It is noticeable that both of these objectives involve evaluative words and concepts. It has become increasingly apparent to the author that the development of sound mathematical judgement is very important to the mathematical modeller. One aspect of this intellectual skill was reported in Clements (1982). In that paper the important role played by the choice of notation in the formulation of mathematical models was developed. Infelicitous choices of notation can have an obscuring effect on the mathematical structure of the model and can greatly hinder the mathematical solution of the model and interpretation of the mathematical results to the real world.

In this chapter, however, the importance of an appropriate choice of level or complexity of model will be considered. In section 2 the definition of an appropriate level of complexity in a mathematical model is further discussed and in section 3 an example, drawn from teaching experience, of a problem which elicited from students models of varying levels of complexity and approximation. In section 4 a general discussion of dynamic system modelling programs and mathematical modelling is given and in section 5 the use of such systems to help students explore different approaches to a mathematical modelling problem is discussed.

2. THE CHOICE OF APPROPRIATE LEVELS OF MODELLING

When a mathematical model of a 'real world' problem is formulated there is almost inevitably a choice to be made concerning which features of the real world are important and must be included in the model and which features are unimportant and may be neglected. Observation of the performance of students learning modelling for the first time suggests that one of their major failings is that they attempt to include all possible effects in their initial model and consequently become submerged in excessive detail. This often leads to failure to make significant progress, discouragement and loss of motivation. On the other hand one of the impressive aspects of the work of very experienced modellers is their almost instinctive grasp of the appropriate level and complexity of model that will reveal most about some physical or organisational problem. The ideal model is one that is sufficiently complex to reveal something significant about the real problem whilst also being sufficiently simple to be mathematically tractable.

Other teachers of mathematical modelling have attempted to address this particular problem. For instance the mathematical modelling methodology adopted by the Open University for the MST204 course, described in Berry & O'Shea (1985), includes the compilation of a 'feature list' of factors that *may* affect the problem and its subsequent refining and pruning to produce a list of features whose effects will be included in the initial model. The pruning step is characterised by Berry and O'Shea as including the making of simplifying assumptions. They also suggest that features pruned from the model at this stage may later be reconsidered as the performance of the initial model is critically examined and the model refined. Evidently Berry and O'Shea apprehend the need to inculcate in students an approach that intrinsically encourages them to make simplifying approximations at an early stage. Their students are thus encouraged by the methodology of modelling which they are taught to try out very simple initial models which they can subsequently improve. By the use of this methodology the modeller should iteratively approach a model that is of an appropriate level to answer the question posed by the 'real world' problem.

Another way in which inexperienced modellers fail is in not making appropriate approximations in the mathematics of the model. Retaining the full equations may again often lead the modeller to become submerged in excessive detail whilst making appropriate approximations simplifies the mathematics and keeps it tractable.

Thus far the argument developed may be seen to be equating appropriate with simple. This is not, of course, intended to be the case, for an adequate and appropriate model must represent sufficient of the features of the real problem to a sufficient degree of accuracy to yield useful and realistic results. It is observable however that students of mathematics usually need to be encouraged to simplify their models more than they need encouraging to complicate them.

There are thus two criteria for saying that the choice of a level of model for a problem is appropriate; that the model should adequately represent the important features of the real problem and those features only, and that appropriate mathematical simplifications and approximations should be made to the mathematical formulation of the model to render its solution tractable and efficient. These two aspects are not, of course, entirely or necessarily independent. A proper understanding of the physical basis of the problem is necessary if the first criterion is to be met but it must also be realised that the physics of the real problem will often also indicate the appropriate mathematical simplifications. At the same time the mathematical features of a developing model, when related and compared with the physical features of the real problem, sometimes indicate possible areas of misapprehension about the real problem. The resolution of these misapprehensions often lead to the realisation that the mathematical model could be altered or simplified.

3. AN EXAMPLE

As an illustration of the effects of choosing different levels of approximation consider the following example. It is required to calculate the vertical acceleration of an arbitrary point (x, y) on the connecting rod of a reciprocating engine. The general arrangement of the problem is illustrated in Fig. 1. The problem arises in one of the simulation/case studies that the author developed in the work previously mentioned. The problem has been used with student groups over seven years and has elicited a variety of approaches, some more successful than others. The coordinates may be non-dimensionalised by the substitutions

$$Y = y/1, \quad Z = z/1. \quad \eta = r/1.$$

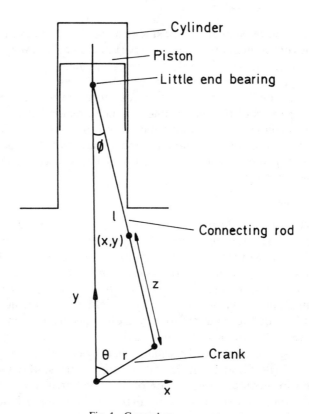

Fig. 1. General arrangement.

Then
$$Y = \eta \cos \theta + Z \cos \phi$$
and
$$\sin \phi = \eta \sin \theta.$$

If the crankshaft turns at a constant angular velocity, w, so that $\theta = wt$, a page or so of tedious manipulation will reveal that

$$\ddot{Y}/\eta w^2 = -\cos \theta - Z\eta(\cos 2\theta + \eta^2\sin^4\theta)/(1 - \eta^2\sin^2\theta)^{3/2} \qquad (1)$$

This expression is exactly correct but somewhat unwieldy. Almost without exception those groups of students who have pursued this avenue have become bogged down in algebra before or, at best, very soon after this point. On the other hand reference to the physical data available for the problem reveals that the parameter η has a value of 0.2. This suggests that terms of the order of η^2 and higher might be ignored. This may be done in two ways. The expression above may be manipulated and the higher order terms removed or, more efficiently, the decision to neglect the higher order terms may be made at the start of the analysis. In the latter case the complexity of the analysis is greatly reduced. In either case the result is

$$\ddot{Y}/\eta w^2 = -\cos \theta - Z\eta \cos 2\theta. \qquad (2)$$

There have been groups of students who have achieved this result by both the routes mentioned. The next part of the analysis was considerably easier for these students and their progress, in general, was better.

The above illustrated an example of an appropriate mathematical approximation, whose validity was suggested by reference to the physical parameters of the real problem, which greatly reduces the complexity of the analysis. On the other hand there was one group who adopted a much simplified model in the first instance. They made the assumption that every point of the connecting rod has the same vertical acceleration as the big end, that is that

$$\ddot{Y}/\eta w^2 = -\cos \theta. \qquad (3)$$

Whilst further progress through the problem was easy for these students their final results were poor. It can be seen, of course, that their model was equivalent to ignoring all terms of order η or greater. In the light of the fuller analysis it is not unreasonable to expect this to introduce considerable errors. The model adopted in this case was a perfectly good initial model but was too crude to yield adequately accurate final results and needed refining before any useful conclusions could be drawn.

In anticipation of the next section we may use the dynamic simulation package to compare the values given by equations (1), (2) and (3) when $Z = 1$ (that is at the little end bearing). This is, of course, the worst case and so provides an indication of the worst error that may be expected by adopting each approximation. Figure 2 shows a block diagram model that may be used to generate the function in (1). A restricted version of this is

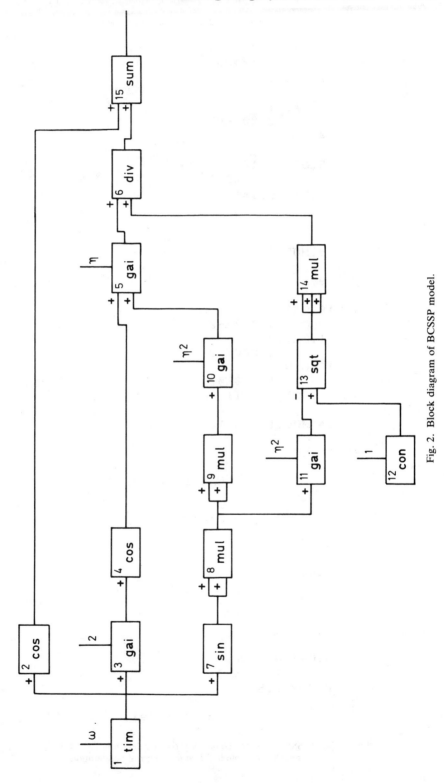

Fig. 2. Block diagram of BCSSP model.

Listing of current system model

1,TIM
      3.142E0
2,COS,1

3,GAI,1
      2.000E0
4,COS,3

5,GAI,4,10
      2.000E-1
6,DIV,5,14

7,SIN,1

8,MUL,7,7

9,MUL,8,8

10,GAI,9
      4.000E-2
11,GAI,8
      4.000E-2
12,CON
      1.000E0
13,SQT,12,-11

14,MUL,13,13,13

15,SUM,2,6

End of model

Listing of current system model

1,TIM
      3.142E0
2,COS,1

3,GAI,1
      2.000E0
4,COS,3

5,GAI,4
      2.000E-1
6,SUM,2,5

End of model

Fig. 3. (a) BCSSP model for connecting rod acceleration. (b) BCSSP model for approximate model of connecting rod acceleration.

used to generate the function in (2). Figures 3(a) and 3(b) show listings of the BCSSP models and in Fig. 4 the results are compared. It is seen that, for this case, the form of (1) and (2) is, for all practical purposes, indistinguishable whilst (3) is evidently a very crude approximation to (1). It is no surprise then that the predictions obtained using this approximation were poor.

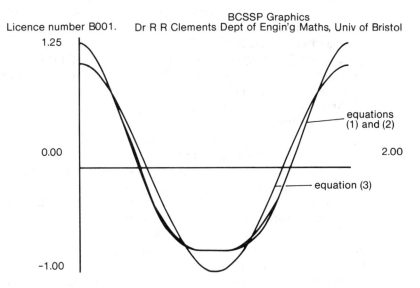

Fig. 4. Approximations to vertical acceleration of little end bearing.

4. DYNAMIC SYSTEM SIMULATION PROGRAMS

The role of simulation in mathematical modelling courses is well established (see for instance Huntley (1984) or Moscardini *et al.* (1984)). Modelling exercises and activities often result in models which lack viable analytical solution techniques. In these circumstances simulation offers a solution route (and one that would be adopted in an industrial or commercial environment). Simulation may be implemented either by a single purpose computer program designed for the problem under study or by the use of one of a range of general purpose simulation systems such as GPSS, CSMP, CSSL, ACSL, DYNAMO, TUTSIM and many others. The major languages like CSMP, ACSL, CSSL, GPSS and DYNAMO are usually available on multi-user mainframe or mini computers. The author's experience of using these languages with students is that they are far from simple to use in the first instance although regular users quickly gain adeptness. Further, most multi-user systems in institutions of higher education are heavily loaded with resulting poor response, at least during class hours. Students who need to simulate a fairly simple system as a part of a modelling exercise usually, in the author's experience, choose to write their own single purpose simulation program in a language with which they are already familiar (usually Pascal, Fortran or Basic). Often they will do this on any microcomputer that happens to be available, particularly as an increasing number own their own microcomputers. The conclusion must be that, whilst simulation should be a regular tool of the mathematical modeller whether student or experienced practitioner, in practice the main available systems are not ideally suited to the needs of tertiary education courses in modelling. This conclusion is reinforced by the experience of Moscardini *et al*. (1984) who describe a simulation package, IPSODE, written within their institution specifically as an introductory simulation package for their students.

Increasingly, however, simulation languages that can be run on microcomputers are appearing. The TUTSIM language, for instance, is a block oriented continuous system simulation language which is available on Apple microcomputers and on a range of other computers running the CPM operating system. It has been used in engineering courses at Bristol and experience indicates that students find it particularly easy to learn and use. The block oriented input language has appeal, particularly to engineering students who are familiar with control theory ideas, in its visual and diagrammatic approach to the construction and representation of models. The package also has particularly good facilities for the graphical presentation of results. In view of these advantages it was desired to make TUTSIM more widely available for use in undergraduate courses in the university. In common with many institutions of higher education, Bristol University has a large number of BBC Microcomputers. The distributors of TUTSIM were unable to provide a version of the language to run on the BBC machine and so a simulation system, BCSSP, which uses largely the same input language and provides similar facilities to TUTSIM was written for the BBC Microcomputer. BCSSP has powerful facilities for the presentation of results in graphical form including the comparison of results from different simulations and can rapidly produce hard copy of graphical output. This facility makes it particularly suitable for its role in the mathematical modelling area. Once available this package opened up possibilities for new teaching styles which are being exploited in the teaching of systems studies, as described in Clements (1985), as well as in teaching mathematical modelling.

5. DYNAMIC SYSTEM SIMULATION AND MODEL COMPLEXITY

The essential point of this chapter is to propose that dynamic system simulation can be used as a valuable tool in encouraging students of mathematical modelling to develop the skill of appropriate choice of model complexity. This is somewhat different from the uses proposed for dynamic system simulation by Huntley and by Moscardini *et al*. (1984). That is not to suggest that their use is in any way invalid; on the contrary it is merely to point out that there is a further role for simulation systems.

Once the students are familiar with the system (and the ease of gaining that familiarity with systems like TUTSIM and BCSSP has been emphasised) they have a powerful tool at their disposal for simulating and examining a wide range of mathematical approaches to given problems. Obviously, in the long term, it is intended that students should be able to choose an appropriate level of model based on their experience and mathematical judgement. On the way to achieving that state of affairs it is desirable that the teacher is able to ask students to explore a range of models of some set of real problems and then to compare the results of these and draw appropriate conclusions from them. Evidently such lessons as may be learned from exercises like this could also be taught didactically but, in keeping with the nature of mathematical modelling as a student-centred activity subject, it may be anticipated that lessons learnt from experience (bitter or otherwise) are much more vivid and are internalised by the learner at a much deeper and more lasting level—that certainly is the author's personal experience.

This suggested to the author that familiarity with a dynamic system simulation program should be taught early in mathematical modelling courses and full use should be made of the system during the course. Such use is both as a simple simulation tool and, because it eliminates much of the labour of setting up simulations of systems, as an exploration tool for the comparison of possible alternative approaches to modelling real problems. Further development of this mode of usage is currently being undertaken.

6. CONCLUSION

In the example given in section 3 it was seen that there was more than one way of modelling the physical system involved and that at least one possible model could be simplified by mathematical approximations. It was also suggested that one of the possible models, suitably approximated, was more useful and valuable for the solution of the real problem than the others. The skill of distinguishing the most useful model of a real problem is a necessary and useful one that teachers of modelling must try to communicate to and develop in their students. A suitably structured component of a mathematic modelling course in which students explore a range of models of some problem areas will help to develop these skills. A readily available and easy-to-use dynamic system simulation program is an invaluable aid to such exploratory work. The BCSSP system has been developed for this and other uses in courses at Bristol University.

REFERENCES

BCSSP (1985). User manual and software package available from Micropacs, Centre for Advanced Technology, Chilworth Manor, Southampton, SO9 1XB.
Berry, J. S. & O'Shea, T. (1985). *Project Guide* (2nd edn), from course materials for MST204 (Mathematical models and methods, Open University Press.
Clements, L. S. & Clements, R. R. (1978). The objectives and creation of a course of simulations/case studies for the teaching of Engineering Mathematics, *Int. J. Math. Educ. Sci. Technol.*, **9**, 97.

Clements, R. R. (1978). The role of simulations/case studies in teaching the practical application of mathematics, *Bull. IMA*, **14**, 295.

Clements, R. R. (1982). On the role of notation in the formulation of mathematical models, *Int. J. Math. Educ. Sci. Technol.*, **13**, 543.

Clements, R. R. (1985). The role of system simulation programs in teaching applicable mathematics. Presented at the 2nd SEFI European Seminar on Mathematics in Engineering Education, DTH Lyngby, March 1985.

Huntley, I. D. (1984). Simulation—its role in a modelling course, In *Teaching and Applying Mathematical Modelling*, ed. Berry, J. S. *et al*. Ellis Horwood, pp. 306–315.

Moscardini, A. O., Cross, M. & Prior, D. E. (1984). On the use of simulation software in higher education courses. In *Teaching and Applying Mathematical Modelling*, ed Berry, J. S. *et al*. Ellis Horwood, pp. 339–355.

20

Discrete and Continuous System Modelling with a Micro Network

I.C. Hendry

and

I. G. Mackenzie
Robert Gordon Institute of Technology, Aberdeen, UK

SUMMARY

Two computer packages used for teaching modelling to undergraduate students are described. They are run on an ECONET network of BBC microcomputers.

The first package uses discrete event simulation techniques to model population growth. The user is presented with a wide choice of deterministic/stochastic models from which to select a model solution. BBC graphics are then used to depict the predicted mean population growth and the range of population growth for the selected model. After comparison between predicted and actual known growth the user may select a better model. The 'modelling cycle' is continued by the student until an acceptable model is found.

The second package is used for solving models of continuous time-dependent systems with known initial conditions. The models which can be studied are those involving ordinary differential equations and the solution of simultaneous ODEs is possible. The method of specifying the equations using the ANALOGUE approach is described showing how students can vary parameters as necessary to obtain a thorough understanding of the model.

The packages have been usd for students in second and third years of a degree course for BSc. in mathematical sciences and for final year BSc. engineering students.

1. POPULATION GROWTH MODELLING

The package for Population Growth Modelling is written as a teaching aid, to be used on a BBC model 'B' computer and disk-drive, not as a comprehensive population growth simulation package. The models which may be simulated are defined by selecting one from a range of birth processes, death processes, and 'litter' size, as listed below.

Birth processes:

(1) Birth-rate fixed (mean time between birth events is constant, regardless of the size of the population).
(2) Birth-rate proportional to population size (mean time between birth events is inversely proportional to the size of the population).
(3) Synchronous births, rate fixed (all cells split simultaneously, mean time between birth events constant).
(4) Synchronous births, birth-rate proportional to population size.

Note that option (4) above occurs as a result of programming logic and does not correspond to any known biological birth process.

Death processes:

(1) Death-rate fixed (mean time between deaths is constant).
(2) Death-rate proportional to population size (mean time between deaths is inversely proportional to population size).

'Litter' size:

(1) Fixed (for synchronous populations this is the only option allowed).
(2) Random size, up to a finite maximum input by the user.

In addition, the birth or death process can be specified, separately to be either random (i.e. Poisson process) or deterministic. Since the number of possible models is very large, and to increase the 'user-friendliness' of the package, the input of a particular population growth model is by a 'conversational' mode of input whereby the user is led through a series of choices in a natural way. An example of the input, for a random birth–death process in which a 'litter', of a random size up to five offspring, is produced at each birth event, is given in Fig. 1. It will be noted that the user is presented with a page of text outlining briefly the types of model available, before being asked to input details of his/her model.

The assumptions on which the simulation of the above models is based are:

(1) The descendants of one parent are independent of the descendants of any other parent.
(2) Only one event (birth or death) can take place at any one time.
(3) The random birth–death process is a Markov process with exponentially distributed times between events. If one process (birth

```
** Econet ** Station  15 **
RUN
IN  THIS  POPULATION GROWTH SIMULATION .NEW CELLS ARE CREATED BY THE SPLITTING
OF  A PARENT CELL TO  FORM ONE OR MORE   ADDITIONAL CELLS. THE NUMBER OF NEW
CELLS CREATED AT A 'SPLITTING' IS CALLED  A 'LITTER':
THE SIZE  OF A 'LITTER' MAY BE FIXED.OR IT MAY TAKE  ONE OF A RANGE  OF VALUES.
WITH FIXED PROBABILITIES.THE PARENT CELL DOES NOT DIE AFTER SPLITTING.
IF ALL CELLS SPLIT SIMULTANEOUSLY      ( SYNCHRONOUSLY). THEN IT IS ASSUMED THAT
  THE NUMBER OF CELLS IN A 'LITTER' IS FIXED
THE POPULATION DECREASES BY THE INDIVIDUAL DEATHS OF CELLS
INPUT INITIAL SIZE OF POPULATION
?10
DO ALL CELLS SPLIT SIMULTANEOUSLY?
IF SO. ENTER NUMBER OF OFFSPRING.IF NOT ENTER 0
?0
IF NUMBER OFFSPRING FIXED. ENTER 1.ELSE 0
?0
WHAT IS THE MAXIMUM NUMBER OF OFFSPRING IN ONE LITTER?
?4
PROBABILITY OF                1  OFFSPRING IN THE LITTER IS?
?0.4
PROBABILITY OF                2  OFFSPRING IN THE LITTER IS?
?0.3
PROBABILITY OF                3  OFFSPRING IN THE LITTER IS?
?0.2
PROBABILITY OF                4  OFFSPRING IN THE LITTER IS?
?0.1
IS BIRTH RATE FIXED OR IS IT PROPORTIONAL TO POPULATION SIZE-ENTER 1(FIXED) OR 0
(PROPORTIONAL)
?0
DITTO FOR DEATH RATE-ENTER 0 OR 1
?0
INPUT BIRTHRATE.DEATHRATE & RUN-TIME(TIME MEASURED IN MINUTES. RUN-TIME<50.0)
?0.01.0.02.30.0
IS BIRTH PROCESS RANDOM?
ENTER 1(YES)0(NO)
?1
DITTO FOR DEATH PROCESS
?1
ENTER NUMBER OF TIMES POPULATION IS TO BE SIMULATED(>3)
?10
IF FOOD SUPPLY LIMITED.ENTER CONSUMPTION/CELL/MINUTE AND FOOD LIMIT. ELSE ENTER
-1.0.-1.0
?0.01.1000000.0
IF YOU REQUIRE SEMI-LOG GRAPHS OF POPULATION SIZE.ENTER 1.ELSE ENTER 0
?1

Not found at line 580
```

Fig. 1. Sample input for population growth simulation.

or death) is deterministic then the other is simulated as a Poisson process with a time-varying mean (Klein & Roberts, 1984).

These assumptions are similar to those used in the simulation of simple random birth–death processes, corresponding to cell population growth models where each cell splits into two or more new cells at randomly distributed times (Jagers, 1975; Bartlett, 1978).

In addition to details on population growth, the user may specify a 'food -consumption' rate, and the amount of 'food' available. It is assumed that each member of the population consumes 'food' at the same rate, regardless of the size of the population, until all the food is consumed.

Output from the package consists solely of graphs of maximum, minimum and man population sizes, computed from a series of at least four simulations of the population growth model, and a graph of the average 'food' consumption of the simulated populations, and minimum sizes of the series of simulated populations, at each time plotted.

Although it would clearly be useful to output more information, such as the maximum and minimum 'food' consumption, or some sample population growth curves, there is insufficient storage on the BBC computer to hold the graph-drawing program and additional information for more graphs. Also, it was felt that more than four graphs displayed simultaneously would be too confusing for the user.

The user may select either Cartesian coordinates, or semi-log graphical output. The graphs are automatically scaled to make full use of the display screen.

Examples of graphical output are given in Figs 2 and 3.

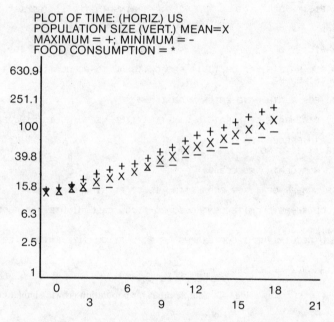

Fig. 2. Semi-log graphs of population growth for a pure birth process.

By means of a series of graded tutorial exercises, students can use this package to discover properties of some population growth models, and of similar stochastic processes, *inter alia*

(a) Exponential growth/decay of simple birth–death processes, whether deterministic or stochastic.
(b) The relationship between initial population size and the probability of extinction.
(c) Compare the properties of deterministic models with stochastic models, particularly with regard to population extinction.
(d) Discover the effect random 'litter' size has on the range/fluctuation of population size.

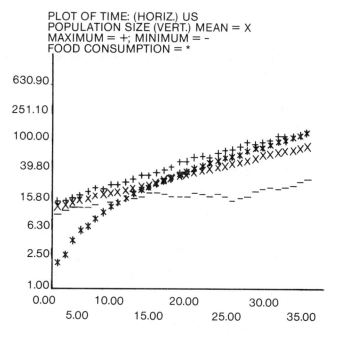

Fig. 3. Semi-log graphs of population growth for a birth–death process.

Once a student has become familiar with the use of the package, and/or the properties of birth–death processes, the package can be used in conjunction with other modelling tools, in a modelling exercise. As an example, consider the following situation.

A biologist has a requirement to rapidly assess the density (number/cubic centimetre) of live bacteria in a culture. It is possible to rapidly assess the density of all bacteria (alive or dead) in the culture by optometric techniques, and it is also possible to measure the amount of 'food' consumed by chemical analysis, at any time, but it is not possible for the biologist, with the equipment at his/her disposal, to rapidly assess the density or proportion of live bacteria.

Experimental data on the numbers of all bacteria, and the amount of 'food' consumed, is available (Figs 4 and 5).

A path to the 'solution' of this modelling exercise could be:

(1) Try, by drawing a straight line by eye, or by using linear regression techniques, to fit a straight line to the experimental data, plotted on semi-log graph paper. Observe that the data fluctuates randomly about the fitted straight line, so that it is unlikely that the process is deterministic.
(2) Using the population growth simulation package, obtain a graph of mean population growth for a simple birth process (no deaths) with birth-rate proportional to population size, which matches the graph of total bacteria numbers.

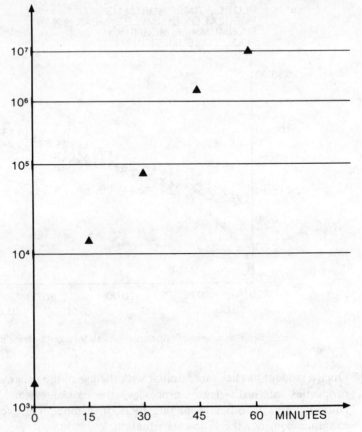

Fig. 4. Bacterial cell numbers (live or dead) versus time.

(3) Discover that it is not possible to find a 'food' consumption rate which matches the experimental data, for the model in (2).
(4) Show that the mean population for a birth–death process, with rates proportional to population size, is given by

$$M(t) = N_0 \exp[(a-b)t]$$

N_0 = initial population, a = birthrate, b = deathrate, t = time either by using the theory of stochastic processes, or by experimenting with the population growth simulation package. Note also that semi-log graphs of 'food' consumed and population size, plotted against time, are asymptotically approximately parallel. Hence the value of $(a-b)$ can be estimated from the slope of the semi-log graph of the experimental data of 'food' consumed.
(5) Fit different birth–death models, subject to the constraint that the difference between birth and death rates is constant. It is found that a good fit to the experimenal data is found when $a = 0.1$ and $b = 0.05$ (see Fig. 3 for a graph of the simulation output).

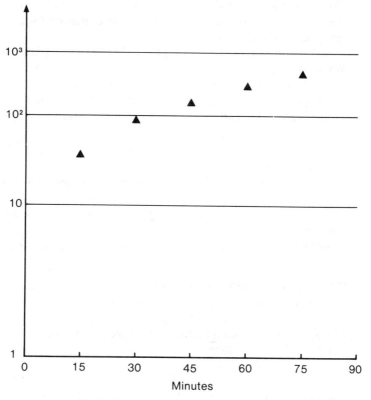

Fig. 5. Bacterial food consumption versus time.

Validation: The biologist took four small samples from the culture after 12 hours, and determined the proportion of live to total cells by staining the samples and examining them under a microscope. The proportions of live cells were found to be 0.5, 0.55, 0.51, 0.42. As these values were within the range of values predicted by simulations of the model found above, in (5), it was concluded by the biologist that the model was satisfactory.

Such modelling exercises, in which the student can experiment with a package to discover the properties of the population, and thereby gain insight into the behaviour of the population, have proved valuable in enhancing the student's modelling expertise, and his/her appreciation of modelling philosophy.

2. CSMP

The second package is CSMP (Continuous System Modelling Package) and like the first, provides a medium whereby students can study many aspects of a given model by varying the parameters; alteration of the model is also possible. It calls for the usual actions on the part of the user, namely:

(1) Specify model (use single ODE or simultaneous ODEs).
(2) Select parameters.

The CSMP 'menu' enables the user to specify the length of time over which the solution is to be illustrated and the time interval for any tabulation or graphical presentation.

The user has a choice of output: either (a) in *graphical* form or (b) in *tabular* form.

The CSMP method of equation solving is easy and parameter surveys can be accomplished quickly as can model improvements. Mathematical modelling can be a stimulating activity when backed up by CSMP.

2.1 Package use

The package is a simulation of an analogue computer using a digital computer. The analogue devices are available for use are: a constant, a divider, an exponential function, a gain, an integrator, a multiplier and a summer.

Some skill has to be developed in representing the equations in block form using the available devices. The steps are:

(a) Write the equation(s) with only the highest differential on LHS.
(b) Create a suitable analogue of the equation(s).
(c) Write the configuration—specify the linking of devices.
(d) Write the parameter specification for the devices.
(e) Use CSMP package.

2.2 Illustration

Students can be directed towards studying S-shaped curves which can represent the diffusion throughout society of some new invention or innovation. If the diffusion process involves the spread of awareness through 'keeping up with' one's neighbours then the rate of acquisition, dy/dt say, might be proportional to y, where y = number of people who have already acquired; so we write

$$\frac{dy}{dt} = ky \tag{1}$$

but quickly realise that there must be a limit to y so equation (1) needs amplifying; it is then suggested that, at time t, the rate of diffusion will depend on the number of potential users still left in society, i.e. $(y(\infty)-y)$, where $y(\infty)$ is the ultimate number of users. Thus a more acceptable model for study is:

$$\frac{dy}{dt} = ky(y(\infty) - y) \tag{2}$$

This model is then studied using steps (a) to (e) above as follows:

Step (a) The equation is already in suitable form.

Step (b) It is necessary to select an integrator as the central block; see Fig. 6 for the salient features.

The integrator has input $\dot y$ and output y and requires the value of $y(0)$ to be specified.

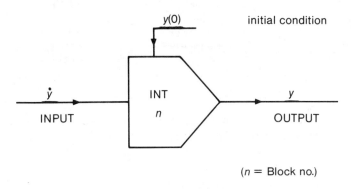

Fig. 6. Block diagram of integrator.

The available analogue devices have to be 'attached' to the integrator to complete the analogue representation of the equation as shown in Fig. 7. At this stage the block identifiers are being used as in the CSMP package.

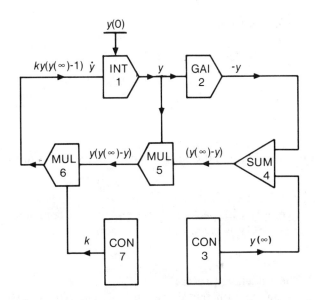

Fig. 7. Block diagram for the equation.

258   Mathematical Modelling—Methodology, Models and Micros

The blocks available are:

Block type	CSMP description
Constant	CON
Divider	DIV
Exponential function	EXP
Gain	GAI
Integrator	INT
Multiplier	MUL
Summer	SUM

Step (c) The blocks are then numbered individually, e.g. from 1 to 7 in Fig. 7, and the configuration of the model can readily be specified as follows:

BLOCK ; TYPE ; INPUT1 ; INPUT2

```
1   ;INT   ;6               (INT connected to Block 6)
2   ;GAI   ;1               (GAI connected to Block 1)
3   ;CON                    (constant)
4   ;SUM   ;3        ;2     (SUM connected to Blocks 3 and 2)
5   ;MUL   ;1        ;4     (MUL connected to Blocks 1 and 4)
6   ;MUL   ;5        ;7     (MUL connected to Blocks 5 and 7)
7   ;CON                    (constant)
```

Step (d) The parameters then require specifying ; we use:

BLOCK ; PAR1 ; PAR2

```
1  ;  0.10      (y(0) = 0.10)
2  ;−1.00       (output = −1.0 times input)
3  ;  5.00      (constant = 5.0)
6  ;  0.04      (constant = 0.04 = k)
```

(note blocks 4, 5 and 6 do not have parameters requiring specification)

Step (e) To use CSMP it will be found that the layout of step (c) and step (d) comply with package requirements. (A complete illustration of the use of CSMP is given in the Appendix.)

2.3 The modelling process

The parameters for the above example use $y(0) = 0.10$ and $y(\infty) = 5.00$ which might represent, say, the number of users of pocket television sets in the UK (in millions). Students would be encouraged to vary the parameters and even change the model. A model change sometimes suggested is to add an advertising factor using

$$\frac{dy}{dt} = ky(y(\infty) - y) + a(y(\infty) - y)$$

where a is the parameter controlling effect of advertising.

2.4 General comments
In continuous system modelling the important activities are:

(1) Agree a model.
(2) Formulate the relevant differential equation(s).
(3) Solve the equations.

Before CSMP was available on a digital computer modelling studies were tedious. Activities (1) and (2) could be stimulating but (3) was time consuming to the extent that the total modelling process lost its attraction. Now, by using CSMP, students can become confident modellers.

The package described was created by a third year student studying for a BSc. in Mathematical Sciences (Hardy, 1984).

He used as a guide his experience with a more elaborate package, CSMP 10, available on the DEC 20 mainframe computer at RGIT. The CSMP 10 users guide, Carnegie-Mellon University, 1971, is part of the DECUS library. Some guidance was also available from Boon (1983).

A full spectrum of models exists in the literature. Some of the books used by the authors are: Doebelin (1972); Jagers (1975); Bartlett (1978); Braun *et al.* (1983) and Klein & Roberts (1984).

REFERENCES
Bartlett, M. S. (1978). *An Introduction to Stochastic Processes with Special Reference to Methods and Applications*, 3rd ed. Cambridge University Press.
Boon, K. (1983). *Explorer's Guide to The ZX Spectrum and ZX81*. Addison-Wesley.
Braun, M., Coleman, S. C. & Drew, D. A. (1983). *Simulation with Differential Equation Models*. Springer-Verlag.
Doebelin, E. O. (1972). *System Dynamics Modelling and Response*. Merrill.
Hardy, M. A. J. (1984). Simulation of an Analogue Computer on the BBC Microcomputer, RGIT Project Report.
Jagers, P. (1975). *Branching Processes with Biological Applications*. Wiley.
Klein, R. W. & Roberts, S. D. (1984). A time-varying Poisson arrival process generator, *Simulation*, **43**, no. 4, 193.

APPENDIX: GUIDE TO CSMP USING THE BBC MICRO

Using CSMP
Stage 1. The program requests the configuration in the form:

```
        BLOCK ; TYPE ; INPUT 1 ; INPUT 2
          ?.
          ?  1  ; INT   ; 2           etc. as appropriate
so proceed   (finish with (RETURN))
```

Stage 2. The program the requests the PARAMETERS in the form:

> BLOCK ; PAR 1 ; PAR 2
> ?

so proceed ? 1 ; 3.0   (giving BLOCK 1 output the initial value 3.0)

    etc.

(finish with (RETURN))

Stage 3. The program asks the user to state for which blocks, up to a maximum of 5, he requires the calculated output. The layout is:

> BLOCK A ; BLOCK B ; —— ; —— ; BLOCK E
> ?

so proceed ? 1   (thus requesting the output from BLOCK 1)

Stage 4. The program asks the user to state the minimum and maximum values to be handled in graph plotting. The layout is:

> MINIMUM ; MAXIMUM
> ?

so proceed ? 0 ; 20   (if the range 0 to 20 is required)

Stage 5. The program asks for time increment and duration in the form:

> TIME INCREMENTS ; DURATION
> ?

so proceed ? 0.1 ; 1.5   (this will ensure calculations are performed for times 0 to 1.5 in steps of 0.1)

Stage 6. This is the option stage which, in time, returns to the screen after every selection of an option.

21

Formulation in Mathematical Modelling by Artificial Intelligence

F. R. Hickman
South Bank Polytechnic, London UK

SUMMARY

It is now well established that the formulation stage of mathematical modelling represents the 'bottleneck' stage of the modelling process as a whole. Until some 'tools' are provided that will assist in unplugging this blockage there seems little hope in establishing a realistic methodology that will underpin teaching and learning strategies. Whether we are teaching mathematical modelling as a vehicle for other disciplines or for its own sake an ultimate goal for the teacher must be to teach others 'how to model'.

The task is a daunting one. We must unravel, in some way, the thought processes of expert modellers. If we can understand the cognitive structures that underly formulation, then and only then, we will be some way to establishing a set of heuristics that might form the basis of a theory of instruction and perhaps even a model of learning.

In its endeavour to understand some of the principal cognitive processes that lead to productive human thinking, educational psychology has turned to the field of artificial intelligence for inspiration. Many of the actions and internal workings of the modern computer seem to simulate human brain function in a remarkable way. Work into long-term and short-term memory storage and retrieval have been particularly productive. This new field of research, called cognitive science shows much promise and in fact gives us the opportunity, as never before, to understand human reasoning.

The last decade has seen a shift in emphasis in the approach of those researchers in artificial intelligence active in this area. No longer is the power of the computer seen as its main virtue but rather its ability to store and manage vast amounts of data (knowledge), i.e. the approach has become knowledge driven. Central to this change in approach has been the development of 'expert systems'. As their name implies these systems aim to act like experts and so are able to solve complex problems in varying domains of activity. In some areas such as medical diagnosis they have been particularly successful.

It is the aim of this chapter to show how the results of expert system design and technology can help us in establishing those heuristics spoken of earlier.

1. INTRODUCTION

It is interesting to note that Professor Burkhardt told the conference on Teaching Mathematical Modelling held at Exeter University, England, in the summer of 1983 that he considered 'some modelling skill and understanding of the processes involved are probably an important ingredient in teaching it successfully' (Burkhardt, 1984). At that time I considered that utterance a sound and valid statement. Today it appears almost prophetic for mathematical modelling at least!

The literature on the teaching of mathematical modelling abounds with examples and case studies. The modelling fraternity are more than willing to share their experiences and in this way a collective 'folklore' of knowledge and practice has built up. However, this is not good enough. There must be some established theoretical framework to guide teachers. After all there is a long standing tradition in education that the fundamental researches in educational psychology are eventually translated into theories of instruction. Teachers are notoriously slow at introducing new ideas into their classroom or lecture theatre and it is to the credit of the educators present and past that mathematical modelling has so quickly established itself as a valid method of instruction.

Traditional theories of learning and teaching have relied heavily on the science of psychology—itself a relatively young discipline and indeed herein perhaps lies one of the major problems in education today. The present lack of unanimity in theoretical approach makes the choosing (by the teachers) of one theory over another an ill-advised thing to do. None of the present theories command obvious priority over the others, nor can a single theoretical formulation, as presently developed, be considered adequate to cope with the diversity of situations arising in the contact with pupils and students. Teachers have therefore tended to remain aloof and impervious to the findings of their colleagues preferring to rely on that folklore and personal experiences previously mentioned. This is a pity, though understandable. The vast amount of psychological material bearing on the work of the teacher with its numerous gaps, inconsistencies, and the possibility of alternative interpretations together with the fact that even rigorous research findings do not in general lead directly to an unequivocal pedagogic prescription make the task of the conscientious teacher a difficult one.

One way around this problem is to provide the teacher with a set of heuristics or rules of thumb that characterise a process of reasoning that itself is psychological in nature. While heuristics are not as elegant as a complete theory they are of immense practical value to the teacher and the pupils. Humans are in fact vast store-houses of heuristic systems for handling information and therefore it would appear natural that any reasoning processes that are emphasised in learning and teaching situations should take these natural heuristics into consideration. This chapter describes how modern approaches to educational psychology have used the computer to simulate these reasoning processes giving rise to the birth of a new discipline—cognitive science—that deals with the interaction between psychology and artificial intelligences; one learning from the other.

Methodologies of mathematical modelling tend to concentrate on taxonomies that describe the process as a whole and few if any give more than passing reference to the formulation stage of the process. They may acknowledge that formulation is difficult, that it represents the 'bottleneck' stage of the process but little work has been done especially in this area to remove that blockage (Treilibs, 1979; Oke, 1984; Hickman, 1985). The processes that are involved in the formulation stage of mathematical modelling are undoubtedly complex. They are closely related to the processes involved in what has become known as problem solving, they are, however, different in nature. This chapter attempts to establish a methodology using the techniques of artificial intelligence and cognitive science to unravel these processes and thereby present a set of heuristics for use by the teacher and the student in their modelling activities.

2. PSYCHOLOGY: LEARNING AND TEACHING (Mouly, 1973)

Psychology is the science that is perhaps most directly concerned with the study of behaviour. In its broadest sense, educational psychology is concerned with the application of the principles, techniques, and other resources of psychology to the solutions of the problems confronting the teacher as she/he attempts to impart the process of 'education' to the pupil. What we are searching for here is information, theoretical or experimental that will help us understand the formulation process. In fact we require more. If we can establish an understanding we want to initiate a set of heuristics that will aid both learning and teaching. These heuristics have a very definite function to perform. They must assist the teacher and the pupil to stimulate, guide and generally facilitate the pupils' learning so as to achieve a set of pre-defined and meaningful goals. Because of the very nature of the formulation process the teacher–pupil relationship must be one of mutual interaction. In fact the teacher should play a more passive role than is perhaps the case today.

There are two main strands to educational psychology—theoretical and experimental. The attempts to provide a theoretical perspective for the empirical findings of modern psychology have led to the formulation of a number of competing theories, some of which have a substantial influence on modern educational attitudes. This is particularly the case in mathematics education. Generally, contemporary theories of learning can be classified into two major systems: associative theories and field or cognitive theories. It is important to emphasise that these theories are really 'theories of behaviour, i.e. viewpoints from which the empirical data of psychology are structured into theoretical perspective.

2.1 Associationism

Associationism dates back to Aristotle's concept of the association of ideas based on similarity, contrast and contiguity. Early psychologists used introspection as their standard approach and rapidly realised that the subjectiveness of this approach was not 'scientific'. As a reaction psychologists and in particular Watson turned to the study of overt behaviour, founding the school now known collectively as behaviourism. Today the behaviourist school manifests itself in two main streams; connectionism (the so-called Bond or S–R theory) and conditioning.

The patriarch of connectionism was Edward L. Thorndike. His theory rests on the following conjecture:

> When a modifiable connection between a situation and a response is made and is accompanied or followed by a satisfying state of affairs, that connection's strength is increased: when made and accompanied or followed by an annoying state of affairs, its strength is decreased' (Thorndike, 1913).

This is the famous Law of Effect. If as Thorndike suggested, bonds were created by repeated pairing of stimuli and responses, then it seemed the teacher's job was merely to provide the proper amount of practice, in the proper order, on each class of problems. It was the teacher's job to identify the bonds that constituted the subject matter concerned, arrange a hierarchy of difficulty (easy first) and then arrange for the pupil to practice each of the kinds of bonds. His work relating to arithmetic was of particular significance (Thorndike, 1922), leading to the familiar practice of drill. An early detractor from Thorndike was William Brownell (Brownell, 1928) who suggested that responses should be meaningful rather than just automatic and if so would lead to greater understanding. We shall see that while Brownell did not present any substantial systematic theoretic justifications for meaningful instruction his influence is still significant today.

Related but not identical to connectionism is the idea of conditioning. Classical conditioning is best represented by Pavlov's well known experiments which relied on the relatively mechanistic process of reinforcement for the formation and strengthening of associations between stimulus and response. Although based on the classical conditioning model, Guthrie's contiguous conditioning theory (Guthrie, 1959) differs from classical conditioning in that it makes contiguity of a given stimulus and response the only factor necessary for the two to become associated. According to Guthrie motivation has no place in his theory and the teacher's role is simply to induce the pupil in any way whatsoever, to make the desired response when the stimulus is present.

Perhaps most complete of all and a paradigm for a model of the systematic and quantitative approach to the psychology of learning is Hall's behaviouristic, reinforcement theory (Hall, 1943). A deductive theory that attempted to derive the laws of learning, it contrasted with Guthrie's theory in that it held the view that reinforcement, in the sense of the reduction of the tension associated with the frustration of drive, is both necessary and sufficient for associations to be formed. Somewhat more complex is instrumental conditioning here because the operant's response is instrumental in accomplishing a given objective, i.e. instrumental conditioning is goal directed. In classical conditioning the behaviour to be conditioned is obtained by some known stimulus (food in the case of Pavlov's dog), whereas in instrumental conditioning the nature of the stimulus is largely irrelevant, one waits for the response to occur spontaneously and then reinforces it. In instrumental conditioning there is a feed-back mechanism that causes the response to undergo continual modification as a consequence of the previous response. In this way there is a gradual shaping of the response towards greater adequacy through successive approximations geared to a schedule of differential reinforcement. It is this latter aspect of instrumental conditioning, which closely simulates desired classroom learning, that prompted the success of Skinner's introduction of programmed learning (Skinner, 1959). Importantly, Skinner has no interest in physiological explanations for behaviour. All that is important is that a rat, pigeon or child learns a certain association, it is not relevant how this is accomplished.

2.2 Field or cognitive theories

The second major family of contemporary learning theories originated around 1912 with Wertheimer's famous gestalt theory. This theory focuses on the global aspects of the situation and is contrary to the mechanistic and atomistic orientation of associationism. The central thesis of gestalt psychology is that thinking and perception are dominated by an innate tendency to apprehend structure. This being the case, the experience of perceiving or thinking achieves an organisation that is more than the sum of objectively identifiable individual elements or stimuli. Theories such as gestalt theory are called field theories because they relate to the whole psychological perspective in which a person operates at any given moment. Learning involves structuring the cognitive field and formulating cognitive patterns corresponding to the relation among stimuli in the environment. The emphasis is on organisation, relationship, meaningfulness and cognitive clarity.

This field approach is emphasised in Lewin's topological theory (Cartwright, 1959), which introduced the notion of life-space, i.e. the psychological world in which the individual lives. Learning, according to Lewin, is a matter of differentiating one's life-space so as to connect more of its subregions by defined paths by perhaps discovering interrelationships among heretofore isolated aspects. Another important concept in Lewin's theory is that of valence, which refers to the strength of attraction and repulsion among the elements of a situation. Also known as the vector theory Lewin postulated that the outcome of a particular situation is the result of the various forces of attraction and repulsion within the overall life-space experienced by the learner. Learning changes the valence value of the components of the life-space. In this way as clarification takes place so a learner modifies the valence of the various goals. By restructuring the life-space so the learner perceives ways that are most likely to enable attainment of these goals.

Tolman's expectancy theory (Tolman, 1959) is somewhat of a halfway house between associationism and field theories in that it is scientifically objective and behaviouristic while, at the same time, emphasising the cognitive nature of experience. According to Tolman, learning involves the establishment of certain relationships between the perception of one stimulus and the perception of another so that a response to a given stimulus leads to the development of certain expectancies. Tolman is strongly against the Law of Effect and the principle of reinforcement. When drives are aroused, the state of tension leads to a demand for goal objects and activity that is guided by expectancies, which the subject attributes to various aspects of the immediate environment. In this way the learner is neither pushed nor pulled by external stimuli but rather follows a path to a goal looking for signs, i.e. learning relationships. In Tolman's theory a correct response confirms the expectancy and therefore increases the likelihood of it happening again, non-confirmation has the opposite effect.

In even sharper contrast to the association theories is the phenomenological version of field psychology (Combs & Snygg, 1959). It represents a systematic attempt to deal with the world of phenomena in the psychological reality of its essential characteristics. It views the individual in a state of dynamic equilibrium within the field of operation placing special emphasis on the phenomenological nature of perception. Perception is defined relativistically; what determines behaviour is not objective but rather phenomenological reality. All the data of experience, that is the total environment is acceptable as the subject matter of inquiry and thus phenomenological psychology is an attempt to throw off the strait-jacket of behaviourism.

2.3 Other theories
There are many other theories that can be mentioned which are of major interest. We may include Gagne's hierarchical model: (Gagne, 1970) in which he lists eight types of learning arranged in a hierarchy ranging from signal learning, that is classical conditioning to problem solving. The distinctive feature of Gagne's learning hierarchy lies in the possible transfer from one level to the next; the learning of principles, for example, involves the chaining of two or more concepts, while problem solving involves the combination of known principles into new elements bearing on a novel problem. We must also mention of course Piaget's cognitive development theory with its four stages of cognitive growth: the

sensorimotor stage (0–2), the preoperational stage (2–7), the concrete operations stage (7–11) and the formal operations stage (11 +). Piaget's approach is of particular interest because of its use of protocol analysis—a technique that will be spoken of later. It must be noted though that the Piagetian model is not without its detractors particularly concerning the variability in children performing Piagetian tasks (Trabasso et al., 1978).

2.4 Implications for teaching and the curriculum in general

It must be recognised at the outset that very few, if any, of these theories of learning, were developed with the problems of the teacher and the pupil as their primary source of inspiration and this is a major criticism. However, it must be a task of the teacher to analyse these theories to see whether they provide a consistent and dependable foundation for educational practice. This is the responsibility of the teachers and their advisers.

A curriculum based on associationism would be characterised by simplicity. In its extreme form it simulates the responses of an educated person by identifying appropriate stimuli and presenting the stimuli in order to match them with the desired responses either through contiguity or reinforcement. This is the basis for the drill programmes mentioned earlier, which have eventually led to many of the CAI programs of more recent years. At a relatively mechanistic level, connectionists would emphasise having the learner primed ready for a given problem with a variable and multiple response and rely on reinforcement to capture correct associations. The desired associations would themselves be consolidated through properly motivated drill and hopefully generalised to related situations. Finally we have the extreme operant conditioning of Skinner that is the basis of programmed learning, so popular not so many years ago. It should be pointed out that for the kind of learning towards which operant conditioning is oriented, Skinner's approach has been eminently successful and can be of great practical value.

A curriculum devised according to cognitive specifications would stress the structuring of the learner's perceptual and cognitive field. The emphasis would be on insight, meaning, organisation and structure. Classroom procedures would be oriented toward the clarification, the discovery of interrelationships and the understanding and implications of structure.

It is clear that learning hierarchies (a la Gagne) have enough psychological reality to justify their use in curriculum design. They can be useful tools to help teachers make explicit their understanding of the organisation of skill learning and the way individuals differ in the extent of their learning. However, teachers must proceed with caution because most hierarchies that have been proposed have no empirical validation and where validation has occurred it has been fraught with difficulties in interpretation and reliability.

2.5 Implications for mathematics education and mathematical modelling

The problem of making learning meaningful has led, particularly in mathematics to the idea that it is the structures of mathematics that are important. In other words a conceptual rather than a computational approach has been advocated in line with the psychological teaching of J. S. Bruner (1966) and therefore with a theoretical basis firmly in the school of the field psychologists. Bruner's theory of the sequence of conceptual development—enactive, iconic, symbolic—and his consequential theory of instruction raises several important issues regarding the nature of cognitive representations. In the formulation of a mathematical model each of these modes of representation are important. Which is most important? It is not at all obvious that a symbolic representation is more advanced than an iconic representation. In developing a theory of instruction for mathematical modelling we must acknowledge that these fundamental issues need further research. However, the aim here is to indicate a way forward that, while acknowledging these problems, gives a *pragmatic* approach to obtaining that theory of instruction.

Closely related to the structure-orientated approach to curriculum development and to the formulation stage of mathematical modelling has been the interest shown in problem solving and problem solving approaches to teaching. I think it is fair to say that the approach to the teaching of mathematical modelling today relates closely to how problem solving has been taught. Problem solving can be learnt by solving problems and mathematical modelling can be learnt by doing modelling! However, in mathematical modelling we have no clear set of heuristics compatible with Polya's system: Understanding the problem, devising a plan, carrying out the plan and looking back (Polya, 1945).

This system facilitates the discovery of the underlying structure of a problem, i.e. it is an aid to insight, in the gestalt sense. The original gestalt psychologists emphasised the importance of problem structure, however they were not very specific as to the processes involved in the acquisition and understanding of these structures. It was left to one of Wertheimer's students, Karl Dunker (1945) to develop the strategies of problem solving. It was also Dunker who introduced the important notions of top-down and bottom-up processing. The former begins with an analysis of goals and problem reformulation while the latter begins with the analysis of the features of a problem; noting what is present and what can be used.

Can similar strategies and heuristics be developed for formulation in mathematical modelling? Before answering it is perhaps appropriate to lay a distinction between problem solving and formulation in mathematical modelling.

Problem solving, reasoning and thinking are terms used more or less synonymously to refer to a broad variety of complex mediating processes such as the reorganisation of cognitive structure, the elucidation of relationships and correlations, the synthesis of isolated experiences (internal and external). In contrast to learning, which entails the grasping of a fully structured situation, problem solving, for example is a matter of reorganising experience with respect to a problem whose solution is not readily available. Problem solving is characterised by its greater emphasis on flexibility of approach and on insight in discovering meaningful relationships.[1] At the basis of this characterisation is the belief in an underlying mathematical structure. Polya and Wertheimer chose problems of a limited generality. In fact, most of their problems depend on geometry and the ability to obtain some special representation for the formalisation. This limited generality, however, serves as a paradigm for the problem solving strategy.

The formulation of mathematical models is characterised by the lack of underlying structure. Real problems are by their very nature extremely complex and it is the task of the formulator to extract the relevant information by whatever means. The assumption of a pre-supposed conceptual framework at once makes the problem more tractable (Hickman, 1985). To know that Newtonian dynamics, or some aspects of mathematical programming is going to be used establishes the paradigm. Indeed this is the only practical approach possible. When the validation stage of the process reveals that the initial assumptions were inadequate the conceptual framework is the last thing to be changed. The use of, say, non-Newtonian turbulent flow models comes about not through choice but through necessity. It is rare indeed that a real problem is solved in a mathematically elegant way (Einstein's model of universal gravitation is an exception) and economic considerations usually dictate that any formulation that results in an acceptable solution is itself acceptable. In a way this represents the behaviourists view; product rather than process. While this distinction between traditional problem solving and mathematical modelling is clear and necessary there are, nevertheless, obvious areas of overlap. Research into problem solving has told us that in order to develop a successful strategy and hence method of instruction, it is necessary to understand the processes involved in the act of formulation itself. If we can successfully simulate these processes then it will be a natural and desirable step to establish heuristics that will form the basis of our strategy.

Recently psychologists have turned to the computer to help them in identifying and understanding the processes that underly productive thinking.

3. COGNITIVE SCIENCE

Cognitive science includes elements of psychology, computer science, linguistics, philosophy and education, but it is more than the intersection of these disciplines (Bobrow & Collins, 1975). This was how Allan Collins defined a new scientific discipline in 1975. Cognitive science deals with the problem of building an intelligent machine that will simulate human conceptual mechanisms, that is cognitive processes. Because mechanistic approaches based on tight logical systems are inadequate when extended to

---

[1]Meaningful here means that the problem solver identifies problem components and relationships that will lead to a solution of the problem. These are known as means–end relationships.

real-world tasks so the workers in artificial intelligence have tried to become much more psychological. At the same time researchers in psychology have found it instructive to view humans as 'information processors' and have therefore become interested in machine models of real-world knowledge. Thus cognitive science was born!

The relevance of cognitive science to formulation in mathematical modelling is an obvious one. We have already mentioned that the characteristic of the formulation stage of the modelling process is its complexity. The details of the dynamic heuristic processes that govern formulation are enormously complicated. The computer is at least capable of handling some of the details and computer simulation of the human activity can be very rewarding and stimulating. As early as 1962 Simon and Newell showed the validity of this approach (Simon & Newell, 1962). Their program called the General Problem Solver (GPS), enables a computer to solve new, unfamiliar problems in ways that match in a reasonable manner the problem solving behaviour of human beings.

It is not important at this stage to discuss the actual success in problem solving of GPS, what is important is what the model revealed. First a *goal* must be clearly defined and represented. Second, the *present state* or knowledge of the individual must also be represented. Third there must exist a control mechanism that decides whether or not the goal has been achieved. The control tells us where we are now and if that is not the goal then a *discrepancy* exists. The discrepancy is eliminated or at least reduced by making some sort of *operation* which represents the problem solving resources we have at our disposal. The *rules* or conditions of the problem and the objects or *elements* of the problems must also be represented. Finally there must exist a mechanism of *evaluation*.

Also characteristic of the approach is the identification of sub-goals. Simon and Newell illustrate these ideas through a flow chart (Fig. 1) of a problem solving process. The flowchart is held to be representative of the general problem solving format. You have a goal, you are not achieving it; you set up a series of sub-goals to help you achieve it; you institute whatever operations are necessary to achieve your sub-goals; you keep track of where you are in the overall operation by a series of tests until the goal is finally achieved.

What cognitive science is telling us is that humans process *basic programmes* whose components (goals, elements, operations, states of knowledge, rules, tests), have been described previously. Each of us contains a vast supply of these programs and we have been developing them since birth. Piaget calls simple programs *schemas* and more complicated ones *operations*. Luria called them *functional systems*, Miller, Galanter and Pitram call them *plans*. Tolman calls them *cognitive maps*, Bartlett *schemes*, Lewin *life spaces* and Berne *games*. Whatever they are called it is the task here to bring together those basic programs that will be necessary to formulate a mathematical model. The type and nature of the required programs is obviously extremely complex but their number is

Formulation in Mathematical Modelling by Artificial Intelligence

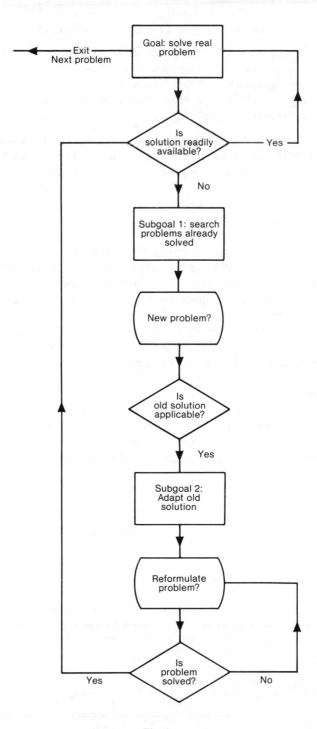

Fig. 1.

surprisingly small. Following Farnham-Diggory (1972) of Carnegie-Mellon basic human information processing needs the following:

a program for scanning and holding information in the mind
a program for solving problems
a program for recalling information
a program for generating and classifying information
a program for ordering and relating information

This is also the case in formulation as the following simple example, taken from the Open University M101 course, shows (Mason, 1982). A boy at the stern of a canal barge leaps off onto the towpath and while the barge keeps moving, runs along the path until he gets to the bow, where he instantly picks up a vacuum flask of coffee and runs back until he gets to the stern. An observer notices that while the boy is running, the barge moved forward a distance equal to its length. How far does the boy run compared with the length of the barge? The solution to this problem is interesting because the author had recorded his attempts at formulating a model (the protocol) of the problem as he saw it. It is of course not the only way to solve the problem but it suffices to illustrate the processes involved. He recognises that the problem is stated at a fairly conceptual level so a physical or enactive model is inappropriate. Here he is using a problem solving program and recalling previous information relating to established modelling methodology. (He is recalling a prototype to be discussed later.) He then draws a diagram (Fig. 2). This iconic stage is a help to short term memory. It is a scanning and holding program because it helps us select the information presented verbally or in print.

It has been shown by George Miller (1969) that we can retain a maxim of only about seven pieces of information concurrently in our short-term memory. For example, a handful of coins cannot be 'counted' at a glance if

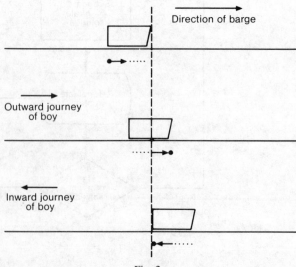

Fig. 2.

it numbers more than about seven coins. In Fig. 2 the author has provided six pieces of information relating to the problem: the direction of travel of the barge, the start, intermediate and finish positions of the boy as the barge travels its own length and the outward and inward directions of the running boy. Again using his problem solving program he begins to make the problem more tractable by making some simplifying assumptions:

Constant speed of barge and boy
Canal is straight

The process of simplification is a process of ordering and relating because he is already relating those aspects of the problem that lend themselves to symbolic representation: speed, distance and time. The generating and classifying program then enables him to recode the information and thus preparing for retrieval from long-term memory. Hence distance = (speed) × (time) and the problem is almost solved, or rather formulated!

This very simple example illustrates the complexity of the cognitive processes involved in the formulation of even the most simple models. If we are to make the assumption that the types of programs just described do indeed function while a human is thinking an important question springs immediately to mind. How do we call up these programs and in what order? This question introduces us to the idea of *control*. It is not known how the human brain controls its own activity and so cognitive science has turned again to the computer to gain insight. In a typical von Neumann type machine control is relatively simple to define. Although the computer may have a large memory store containing large amounts of data and a large collection of commands (or steps in a program) only one step in one program is active at any one instant in time. When this step has been carried out, some other step must be selected to become active. The method by which this next active step is chosen is called control. In *single processing with direct transfer of control* each step in a program is numbered and the program is then executed in numerical order. There are two common exceptions to this: (i) If a program step involves an unconditional transfer command (such as Next in Basic), that program step itself determines which command will become active next; or, (ii) if a program step involves a conditional transfer (IF–THEN command), then some specified test is carried out, and which command becomes active next is determined by the outcome of that test. Since only one command is active at any one time and because the active command always chooses the next command we have control. The memory location where the active control is located is called the control register.

A major alternative to this type of control consists of an arrangement where several or perhaps many control registers exist and several of them are active at the same time. It is thought that this system more typically simulates human control mechanisms and has as its basis the concept of simultaneous processing. The initiation of this type of control strategy is complex and still being researched in computer science itself. One method that is much used in AI is that of pattern recognition. Here each program has attached to it one or more pattern recognisers which continually inspect the activities of the ongoing information processing. Whenever a pattern recogniser recognises its trigger pattern, it causes its attached program to enter one of the unoccupied control registers and hence to become active. Computer simulations of human thinking have been extremely successful using these techniques.

Thus far we have suggested that cognitive science offers us the opportunity of understanding the actual processes that are involved in the formulation of mathematical models. The idea that human thinking can be simulated in this way through identifying *'cognitive programs'* and *'control strategies'* is highly provocative. However, how do we establish the content of these programs? For example, if we wanted to produce a flow-chart for a relationship or rule discovering process that was part of the ordering and relating program, how would we go about it? One way is to build or attempt to build what has become known as an expert system which would automatically utilise those features identified as useful by cognitive science.

4. EXPERT SYSTEMS

There is a lot of misunderstanding surrounding expert systems and artificial intelligence (AI). Expert systems are derived from Artificial Intelligence. AI appeared in its modern form about 1975; I say in its modern form because relatively early in its history there was a shift from a power-based strategy to a knowledge-based approach (Goldstein & Papert, 1977). AI was seen as a scientific subject rather then an engineering discipline and its theatre of application was the study of the principles of intelligence using information processing concepts as its theoretical framework and the computer as its principal tool. The theoretical change in emphasis was brought about by the realisation that human intelligence depends upon the vast range of knowledge that we bring to bear when solving a problem. Thus by about 1975 a number of AI systems had been developed that appeared to solve difficult practical problems in a wide variety of fields or domains. It was not that intelligence as such had been understood but rather that symbolic information processing techniques had been developed which were able to solve specialist problems at a level that seemed comparable to the performance of human experts. These systems have now been called, collectively, expert systems. One of the reasons why expert systems have had a bad press is not because the systems did not work[1] but rather that the subject areas that AI have tackled have usually already been tackled by other disciplines. Hence problem solving and inference have been looked at in a conventional information processing way by decision analysts and adaptive control theorists respectively. However, it is the basis of approach of the AI worker that is different and central to this approach to Expert Systems is the representation and use of

---

[1]One of the earliest systems, MYCIN, has consistently out-performed experts in extended clinical trials at the medical school of Stanford University.

knowledge. There is no one standard definition of an ES but the following captures the spirit of most systems

LOGIC + CONTROL = ALGORITHM

ALGORITHM + KNOWLEDGE = EXPERT SYSTEM

Expert Systems consist of three major components: a knowledge base, an inference engine and a user interface (Fig. 3).

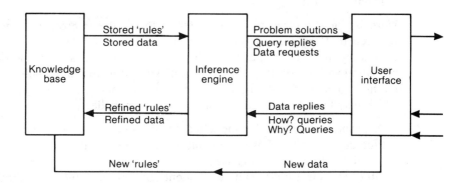

Fig. 3.

The user interface concerns itself with system communication, ideally it will incorporate some form of natural language processor which is able to cope with a number of how? and why? queries expressed in everyday language. The knowledge base is a database containing a very large body of knowledge about a particular domain. Unlike 'normal' information processing systems the knowledge base of an Expert System emphasises qualitative logical reasoning rather than quantitative calculation. Its contents are not abstract symbols like conditional probabilities or other numbers but will typically hold a large number of facts and rules (and sometimes rules alone). In expert systems the aim is to represent meaning explicitly by recording these relationships in a way that reflects *peoples'* conceptualisation of the relationships and rules, while in a form that the computer can exploit rather than reducing them to abstract quantities. The heart of the expert system is the inference engine; the point of the program which embodies the algorithms for eliciting information from the user, searching the knowledge base and generating inferences. Since real-world knowledge frequently involves uncertainty the inference engine must be able to handle uncertainty in arriving at a solution. In particular much work has been done in establishing methods of reasoning from incomplete knowledge (Collins *et al.*, 1975). We will return to this latter in the specific context of mathematical modelling. Inextricably linked with the inference engine is the knowledge representation scheme. The choice of the representation scheme, that is the system into which the knowledge is encoded, for a particular domain is critical (Johnson & Keravnon, 1983). The scheme should accurately capture the underlying conceptual structure of the declarative domain knowledge. There are as many knowledge representation schemes as there are expert systems. The main ones can be categorised as follows:

Predicate calculus
Associative networks
Frames
Production rules

Here is not the place to examine all of these, the interested reader is referred to Barr, Cohen and Feigenbaum (Barr *et al.*, 1981). The most commonly used representation is that of production rules. Their mode of operation within expert systems is best illustrated by formally applying the method to a 'body of knowledge' relevant to mathematical modelling which may have been elicited from an expert modeller.[1]

A problem that can be modelled by the class of models known as time-dependent models often exhibit the features that the variables can be related to *inputs* that cause increase and *outputs* that cause decrease. Other models are also time-dependent but some exhibit the additional feature that there is a delaying factor.

If we now make the assumption that this knowledge can be organised as a set of laws and facts, we can analyse the knowledge as follows:

A: time-dependent models
B: time-dependent models with delay
C: discrete time-dependent models (say)
X: input feature
Y: output feature
Z: time delay feature

Laws:

L_1: any model A exhibits features X and Y
L_2: any model B exhibits feature Z
L_3: a specialisation of a model type inherits the features of that category

Facts:

F_1: B is a specialisation of A
F_2: C is a specialisation of A
F_3: A is a model category for B and C.

In AI terminology the laws L_1 and L_2 represent *declarative* knowledge and the law L_2 represents procedural knowledge. The procedural knowledge dictates the conceptual structure of the declarative knowledge. For

---

[1]The process of elicitation and its importance in the design of expert systems will be discussed later. This particular body of knowledge is totally artificial.

example, according to some procedural knowledge component certain facts should be clustered together and manipulated as a single entry. This body of knowledge can now be encoded in terms of rules. The deductive relationships involved are as follows:

$$D_1: \frac{\text{model is of type A}}{\text{exhibits features X and Y}}$$

$$D_2: \frac{\text{model is of type B}}{\text{exhibits features X,Y and Z}}$$

The knowledge is represented in rule form as the reversion of these deductive relationships:

R_1: *if* the problem exhibits feature X *and* the problem exhibits features Y,

then we may conclude that the likely model is of type A.

R_2: *if* the problem exhibits feature X *and* the problem exhibits feature Y *and* the problem exhibits feature Z,

then we may conclude that the likely model is of type B.

Since the first two clauses of the antecedent of rule R_2 represent the consequent of rule R_1, R_2 may be re-expressed as

R_2: *if* the model is of type A *and* the problem exhibits the feature Z,

then we may conclude that the likely model is of type B.

The knowledge is encoded in this form precisely to infer hypotheses rather than to represent deductive relationships, that is if the antecedent of R_1 is satisfied (i.e. the problem exhibits the feature of input and output variables) then we may infer that we can probably use a time-dependent model. It should be noted that since the knowledge itself embodies uncertainties, the conclusion will be qualified by a degree of confidence (itself inferred from perhaps Bayesian probabilities or the expert himself). Rules such as R_1 and R_2 encode knowledge that is suitable for inference and in this way the system can be made to answer How? or Why? questions simply by displaying the rules.

Other rules which do not constitute reversions but could also be included in the knowledge are:

R_3: *if* the model is of type A *and* the problem *does not* exhibit feature Z,

then we may conclude that the problem is not likely to be described by model B.

R_4: if   the model is of type A *and not* of type B,
      then   we may conclude that the problem could be described by model C.

R_5: if   the model is of type A *and* the model is *not* of type C,
      then   we may conclude that the problem could be described by model B.

One of the defects of this system is that the taxonomy of models (facts F_1, F_2, and F_3) are not explicitly represented through the rules. However, there is an implicit representation in that 'the model is of type A' is in the antecedent of each rule and that model types B and C are mutually exclusive. We can see for example, that the rule R_2 is the reversion of the deductive relation D_2, D_2 in turn represents the law L_2 and the application of the law L_3 to law L_1 and facts F_1 and F_3.

If it is difficult for production systems to capture all the knowledge, then as mentioned previously, there are other representations that can be used. An important area of development in expert systems is the 'object oriented' or frame style of representing concepts as structured collections of interrelated facts and rules. The formalism was first introduced formally by Minsky (1975) although the actual concept can be traced back to Bartlett (1932). A frame is a data-structure consisting of a network of nodes and relations used for representing a stereo-typical situation. Minsky uses the example of a birthday party to illustrate the idea:

> Jane was invited to Jack's birthday party, she wondered if he would like a kite. She went to her room and shook her piggy bank, it made no sound.

Most readers would interpret this story to mean that Jane wants money to buy Jack a present but there is no money in her piggy bank. Minsky and other researchers such as Schank (Schank & Abelson, 1977) suggest that this response is surprising in that the words 'present' and 'money' do not appear in the story, but the connection has been made none the less. They would explain this by saying; something early in the story, as seen, triggered the retrieval of the 'birthday frame' from long-term memory which has a slot in it for 'present' as a default assignment (something that may well be expected at a birthday party). The same reasoning applies to the piggy bank frame with the slot, 'money'. In this way, Minsky argues, inference take place.

Attached to any frame is information about how to use that frame, what to expect to happen and what other frames might be related and under what conditions. The importance of the conception lies in the fact that it allows the co-existence of the declarative knowledge representing some situation and the procedural knowledge that controls it. It does this because a frame contains a large body of richly interconnected information about a single topic organised around typical observations and procedures (in the opinion of the expert!). The top levels of a frame are fixed, and represent things that are always true about the supposed situation. The lower levels have many terminals, called slots, that must be filled by specific instances of data. Each terminal can specify conditions its assignments must meet (in fact the assignments themselves are usually smaller sub-frames). The slots are initially filled with default assignments containing information that hold unless new information displaces them. In terms of cognitive processing, Minsky conjectured that frames were never stored in long-term memory with unassigned terminal values, usually they contain weakly bound default assignments (such as 'present' above). As we react with our environment so we call up these frames from long-term memory and modify the default assignments for a particular situation, i.e. the frames are closely related to our notion of basic human programs or heuristics. The notion of frames produces an important link with the field theories of psychology that have previously been described. For example, Minsky believes that there is a similarity between Piaget's idea of concrete operation and the effects of applying a transformation between frames.

Again to illustrate the idea, although it does not do it justice, the frame representation of our modelling example is given in Fig. 4.

The frame representation scheme captures the given conceptual structure and the taxonomy of models is represented through the 'specialisation' and 'type' slots and the empirical associations between features and models through the 'features' slots. Thus all the laws and facts are explicitly represented. It is most important to note that these frame representations are not limited to classification hierarchies but can represent other aspects of knowledge. Part of hierarchies can describe the

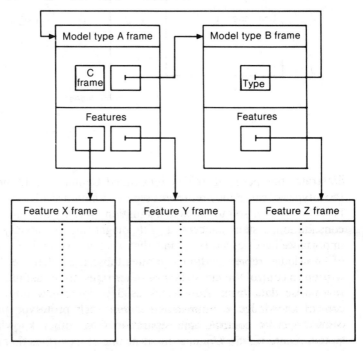

Fig. 4.

frames which are components of other frames and so on, as above. Importantly for mathematical modelling, frames can be more abstract in that they can be used to represent meta-knowledge, i.e. knowledge about knowledge. Systems with meta-knowledge can stand back and critically consider the different techniques available to them in solving a problem. The major step in applying these ideas is to regard all the concepts introduced so far, facts, laws, rules, decisions, control mechanisms as objects themselves, i.e. frames and consequently they can appear in frames themselves. However, Minsky has not made it clear how frame-based systems can infer and it is a feature of modern approaches that frame oriented representations are regarded as being most suitable for representing facts and data-declarative parts of the knowledge. The procedural parts—the parts that control the processes of symbol manipulation (in this case the basic symbol being a frame itself)—are usually dealt with separately. We have already seen that a most convenient way of representing procedural knowledge is through a rule. Rules can be stored within object frames and are used when inferences about an objector set of objects are required. Figure 5 taken from Aikins (1983)

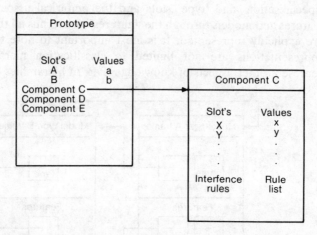

Fig. 5.

illustrates this point. In this system called Centaur the frames are called Prototypes because they represent typical situations which can be used as a basis for comparison to the actual situation given by the data. Centaur is a consultation system concerned with diagnosing pulmonary disease. Its importance here lies not in the fact that it is a consultant but that its method of knowledge representation separates those rules that are, for example, written to control the invocation of other rules, to set default values, or to summarise data from those rules used to infer new information. The control knowledge is represented *within* each prototype thus allowing context-specific control, and separation from other knowledge in the system. Quite rightly Centaur claims as one of its virtues that it represents its knowledge explicitly. Rule-based systems, as we have noted, represent their knowledge implicitly. The rule format represents not only procedural knowledge, but also declarative knowledge and control knowledge within a prototype. If one is using such a system to explain expert performance then there is a significant advantage in Centaur's approach in that experts can specify a different set of control tasks for each prototypical situation.

In mathematical modelling, the formulation stage of the modelling process often calls for the recall of prototypical situations. When I was modelling the rolling resistance of a pneumatic tyre on a road (Hickman, 1985) I was immediately drawn to the classical prototype model that deals with Coulomb friction. This behaviour is typical of the professional modeller particularly where expediency calls for the use of standard techniques, which is very often the case in commercial or industrial environments. The actual formulation and refinement of the model can then take place through the slots of the chosen frame. Where necessary other prototypes may be called up so that the familiar 'stagnation period' is avoided or at least kept to a minimum. There is no doubt that closer examination of the Centaur system will be productive. It is not yet clear, however, that the system would be wholly adaptable to the domains of mathematical modelling because a decision as to the status of the required system (consultative, tutorial etc.) has yet to be made.

It has been mentioned before that the heart of an expert system is the inference engine. All systems have an inferencing component which draws conclusions from the data it is presented with. The method of operation of the inference engine will depend on the type of logic employed, which can be one or more of:

(a) Propositional
(b) Predicts

(i) Boolean
(ii) Extended boolean
(iii) Multi-valued
(iv) Fuzzy
(v) Bayesian
(vi) Probabilistic

Where a degree of uncertainty is to be built into the system care must be taken as to the choice of model for a particular domain in order to avoid odd results. More careful elicitation techniques can sometimes avoid the use of such models.

The above logic systems are only part of the total inference engine. There also has to be a control mechanism or strategy which decides which items of data are to be gathered and used to make an inference. There are many such control strategies, some are listed below:

(a) Blind search
   (i) Breadth first
   (ii) Depth first
(b) Heuristic search
   (i) Best first
   (ii) Generate and test

(c) Directional search
  (i) Backward chaining
  (ii) Forward chaining
  (iii) Bidirectional chaining

The type of search strategy chosen is dependent on the specified domain. For example, in mathematical modelling the system might choose from a list of possible prototypes by a 'best first strategy'. The data drives the selection using a heuristic function or some other routine that decides which prototype looks most likely to lead to a successful solution. Of course cross-checking mechanisms are included that enable a change of prototype as more data is inferred or gathered. Within each prototype object-level inferences can be goal driven (backward chaining) or data-driven (forward chaining). If the control strategy fails to infer a solution the system could ask for more data.

So far, some of the features that go to make up an expert system have been described and their relevance to formulation in mathematical modelling noted. However, having decided that because of its flexibility an object level environment would be a suitable knowledge representation, one might ask the question: where does our control strategy and inferencing logic come from? How do we gather not only the declarative knowledge but also the procedural knowledge specific to formulation in mathematical modelling? The answer is of course from expert modellers!

Most expert systems are rule-based systems. This is because it is a common belief among researchers that most knowledge can be captured in rules. This is a misconception as systems like Centaur have shown. One side-effect of this misconception is that little research has been carried out in the area known as knowledge elicitation,[1] i.e. the extraction of knowledge from an expert. The research that has been done suggests two main methods both of which use the services of the 'knowledge engineer'. The knowledge engineer can be thought of as the project manager acting as a mediator between the expert and the knowledge base. She/he need not be the programmer or gatherer of information but will supervise the tasks that are necessary for successful implementation. According to Feigenbaum (Feigenbaum & McCorduck, 1983) the design task is characterised by rapid prototyping. First the knowledge engineer gets acquainted with the nature and terminology of the domain by studying textbooks, manuals or other written material. Second the knowledge engineer interviews a domain expert and observes him/her solving realistic problems from the domain. The aim of these interviews is the selection of the appropriate tools for knowledge representation and inferencing and the gathering of

---

[1]There is a subtle distinction between knowledge elicitation and knowledge acquisition. Knowledge elicitation denotes the methods and techniques used to extract and explicate the knowledge that drives expert performance. Knowledge elicitation is a subset of the knowledge acquisition stage of expert system design (the term subset is used in its proper form in that in certain circumstances knowledge elicitation is the same as acquisition, i.e. the subset equals the set). Knowledge acquisition in general has a wider scope in that sources of information other than a human expert are consulted, i.e. books, records, users, etc.,

the domain specific knowledge (laws and facts). A first prototype system is then brought up quickly and demonstrated to the expert. A subsequent incremental development of the system follows.

A different approach to the design problem has been proposed by Breuker and Wielinga in a series of reports that form the basis of an Esprit project developed in conjunction with the knowledge-based system centre of the Polytechnic of the South Bank (Breuker & Wielinga, 1983). This ambitious project aims at establishing a methodology for knowledge based system design and the early reports deal with 'the acquisition of expertise'. The main proposal regarding knowledge elicitation is that a greater degree of analysis of the domain is required before the elicitation process takes place. They suggest five levels of knowledge analysis.

Knowledge identification.
This level of analysis corresponds simply to recording what one or more experts report on their knowledge.
Knowledge conceptualisation. Aims at the formalisation of knowledge in terms of conceptual relations, primitive concepts and conceptual models. The knowledge of different experts and possibly different sub-domains is unified within one conceptual framework.
Epistemological analysis. This analysis uncovers the structural properties of the conceptual knowledge, formalised in an epistemological framework.
Logical analysis. This level of analysis applies to the formalisms in which the meta-level knowledge is expressed and is responsible for inferencing.
Implementational analysis. This is an analysis of the control strategies, i.e. the mechanism on which the higher levels of knowledge are based.

Breuker and Wielina's point is that traditional expert system design (Feigenbaum's) map from the first level to the fifth level. They suggest that there is an intermediate stage that will improve the authenticity of system design. This intermediate stage rests on the design of so-called 'interpretation models'. An interpretation model consists of a typology of basic elements, structuring relations and a representation of the inference structure for a class of domains. The elements of the model are canonical, that is, they are abstractions of the elements that constitute the knowledge in a specific domain. A classification of canonical elements has been given by Wielinga (Wielinga & Breuker, 1984). The interpretation model represents a top-down approach to the analysis of verbal data. By some technique structured interview, thinking aloud procedures etc., data is gathered from the expert and matched against the interpretation model; the model is tested and debugged by the 'empirical data'. It is argued that the more elaborate and explicit the interpretation model is from the start, the more efficient the analysis can proceed.

Neither of these approaches are feasible in a domain as complex as formulation in mathematical modelling. How can a knowledge engineer get acquainted with the nature and terminology of formulation? There are no textbooks, manuals and very few research papers (Treilibs, 1979; Oke, 1984). The non-domain expert knowledge engineer has even less of a chance of designing a satisfactory interpretation model. What is required is a synthesis of the two methods, that is a top-down approach (interpretation models) and a bottom up approach (rapid prototyping). The only way that this can be feasible in the domain of mathematical modelling is that an expert modeller plays the role of the knowledge engineer, eliciting knowledge from other experts. The whole exercise is made more tractable in that interpretation models can be constructed by the expert from his own knowledge having predetermined the knowledge representation. This predetermination will significantly simplify the recognition of knowledge sources[1] specific to that domain. The rapid prototyping approach and empirical debugging could then be used for refining the model. Once a finely tuned model has been obtained it can form the basis of the learning and teaching strategies for mathematical modelling.

5. CONCLUSIONS

This chapter has argued that in order to teach modelling effectively we must understand the processes that underly the formulation stage of the modelling process. This stage is well recognised by professional modellers and practising teachers to be the 'bottleneck' stage of the process as a whole. The unplugging of this blockage must be a primary step towards a theory of instruction for mathematical modelling.

The argument for a theoretical description of the processes rests on the traditional approach of educational psychology. Here theoretical descriptions for observed phenomena have given rise to theoretical prescriptions for instruction. Thus we saw that Thorndike's connectionist approach led to the emphasis in the classroom (and lecture theatre!) on drill and practice. Simultaneously, rival field theories eventually led to the structure-oriented approach to curriculum design and the arrival of discovery methods such as problem solving.

The behaviourist school is not interested in underlying mechanisms, giving great emphasis to the products rather than the processes of thinking activity. It is here that the cognitive school departs from the Skinnerian tradition. It must be acknowledged that many of the behaviourists' techniques (Skinner's in particular) are of great practical value (how many of us shut up our children with a bribe of some sort!), but the cognitivists have been unwilling to give up the belief that there is an underlying cause from which the observed behaviours do in fact flow. If this underlying cause, this process, can be understood then it can be taught and the resulting desirable behaviour will follow.

Modern educational psychology has turned to the computer to help in the understanding of some of the fundamental cognitive processes and together with the work resulting from the researchers in artificial intelligence has given rise to the new discipline of cognitive science which considers humans as information processors of one kind or another. The

---

[1] A knowledge source is a piece of knowledge that derives new information from existing data (Clancey, 1983).

success of this approach in explaining observable behaviour has been remarkable (Davis, 1984), and in an attempt to understand the human thought processes cognitive scientists have seen the activities of the researchers in artificial intelligence as a source of inspiration.

The methodologies that exist for building expert systems give us an opportunity to establish our theory of instruction in a pragmatic and practical way. This chapter argued not for the implementation of an expert system in mathematical modelling (though would not that be a desirable thing?) but that the thought that goes into design of such a system can give us the heuristics that would form the basis of a teaching and learning strategy in mathematical modelling. A project designed to implement this argument is already underway.

REFERENCES

Aikins, J. S. (1983). Prototypical knowledge for expert systems. *A.I.*, **20**. No. 2

Barr, A., Cohen, P. R. & Feigenbaum, E. A. (1981, 1982). *Handbook of Artificial Intelligence*. Vols 1 to 3. Los Altos, Calif: Kaufmann.

Bartlett, F. C. (1932). *Remembering: A Study in Experimental and Social Psychology*. Cambridge University Press.

Bobrow, D. G. & Collins, A. (eds) (1975). *Representation and Understanding. Studies in Cognitive Science*. Academic Press.

Breuker, A. J. & Wielinga, B. J. (1983). Analysis Techniques for Knowledge Based Systems. Report 1.2. Part 1 and 2. Esprit Project 12.

Brownell, W. A. (1928). *A Development of Children's Number Ideas in the Primary Grades*. Chicago: University of Chicago.

Bruner, J. A. (1966). *Toward A Theory of Instruction*. Cam. Mass: Harvard University Press.

Burkhardt, H. (1984). Modelling in the classroom—how can we get it to happen? In *Teaching and Applying Mathematical Modelling*, ed. Berry, J. S. *et al.*, Ellis Horwood, pp. 39–47.

Cartwright, D. (1959). Lewinian theory as a systematic framework. In S. Koch (ed.), *Psychology: A Study of Science*, Vol. 2. NY: McGraw-Hill, pp. 7191.

Clancay, W. J. (1983). The epistemology of a rule-based expert system-a framework for explanation, *A.I.*, **20**, 215–251.

Collins, A., Warnock, H., Aiello, N. & Miller, M. L. (1975). Reasoning from incomplete knowledge. In D. G. Bobrow, & Allan Collins (eds), *Representation and Understanding. Studies in Cognitive Science*. Academic Press. pp. 383–415.

Combs, A. W. and Snugg, D. (1959). *Individual Behavior. A Perceptual Approach to Behavior*. New York: Harper & Row.

Davis, R. B. (1984). *Learning Mathematics. The Cognitive Science Approach to Mathematics Education*. Croom Helm.

Dunker, K. (1945). On problem-solving. *Psychological Monographs*, **58** (270), 1–112.

Farnham-Diggory, S. (1972). *Cognitive Processes in Education. A Psychological Preparation for Teaching and Curriculum Development*, Harper & Row.

Feigenaum, E. A. & McCorduck, P. (1983). *The Fifth Generation*. Addison-Wesley.

Gagne, R. M. (1970). *The Conditions of Learning* (2nd edn). New York: Holt, Rinehart & Winston.

Goldstein, I. & Papert, S. (1977). *Cogn. Sci.*, 1.84.

Guthrie, E. R. (1959). Association by contiguity. In S. Koch (ed.) *Psychology: A Study of Science*, Vol. 2. NY: McGraw-Hill, pp. 158–197.

Hickman, F. R. (1985). Didactic and pragmatic approaches to mathematical modelling. *Int. J. Math. Educ. Sci. & Technol*. (Forthcoming).

Hall, C. L. (1943). *Principles of Behavior*, NY: Appleton.

Johnson, L. & Keravnon, E. (1983). The importance of the Knowledge Representation Scheme in the Performance of Expert Systems. MCSG/TR/29. Dept. Compt. Sci., Brunel University. UK.

Mason, J. H., (1982). *Mathematics: A Psychological Perspective*. Oxford University Press.

Miller, G. A. (1969). Some preliminaries to psycholinguistics, *American Psychologist*, **24**, 1063–1074.

Minsky, M. (1975). A framework for representing knowledge. In P. H. Winston (ed), *The Psychology of Computer Vision*. NY: McGraw-Hill, pp. 211–277.

Manly, G. T. (1973). *Psychology for Effective Teaching*, Holt, Rinehart & Winston.

Oke, K. H. (1984). Mathematical Modelling Processes; Implications for Teaching and Learning. PhD thesis. University of Loughborough.

Polya, G. (1945). *How to Solve It: A New Aspect of Mathematical Method*. Princeton University Press.

Schank, R. C. & Abelson, R. P. (1977). *Scripts, Plans, Goals and Understanding. An inquiry into Human Knowledge Structures*. New Jersey: L.E.A. Publishers.

Simon, H. A. & Newell, A. (1982). Computer simulation of human thinking and problem solving. In W. Kessen & C. Kahlman (eds), *Thought in the Young Child, Monographs of the Society for Research in Child Development*, Serial No. 83, 137–150.

Skinner, B. F. (1959). A case history in scientific method. In S. Koch (ed.), *Psychology: A Study of Science*, Vol. 2. NY: McGraw-Hill, pp. 359–379.

Thorndike, E. L. (1913). *Educational Psychology*, Vol. 11, *The Psychology of Learning*. NY Teachers College, Columbia University.

Thorndike, E. L. (19922). *The Psychology of Arithmetic*. NY: Macmillan.

Tolman, E. C. (1959). Principles of purposive behavior. In S. Koch (ed.), *Psychology. A Study of Science*, Vol. 2. NY, McGraw-Hill, pp. 92–157.

Trabasso, T., Isen, A. M., Doleski, P., McLanalan, A. G., Riley, C. A. & Tucker, T. (1978). How do Children solve class-inclusion problems? In R. Siegler (ed.), *Children's Thinking: What Developments?* Hillsdale, NJ: Lawrence Erlbaum Ass.

Treilibs, V. (1979). Formulation Processes in Mathematical Modelling. M. Phil thesis, University of Nottingham.

Wielinga, B. J. & Breuker, A. J. (1984). Interpretation of Verbal Data for Knowledge Acquisition. Report 1.4. Esprit Project 12.

22

Using MODISTAT—Micro Stats Package for Illustrating Mathematical Principles

T. H. Mangles
Plymouth Polytechnic, UK

SUMMARY

Modistat was developed primarily as a tool for statistical analysis. From its conception several criteria were identified as being fundamental.

- the package should be user friendly,
- the package should be easy to use,
- the user prompts should be in English,
- the user should not be left with a blank uninformative screen,
- extensive use should be made of the colour graphics capabilities of microcomputers.

During its development several features were seen to have applications to the teaching of mathematical principles:
This chapter develops some of these applications including:

- the solution of linear equations and consistency
- the solution of equations and the remainder theorem
- the convergence of Fourier series
- discontinuities in functions.

1. SOLUTION OF SIMULTANEOUS EQUATIONS

In statistics we are often interested in predicting a variable, referred to as the dependent variable from one or more other variables, referred to as independent variables. In many cases a linear model is assumed and in one of its simplest forms we have

$$Y = \beta_1 x_1 + \beta_2 x_2 + e \qquad (1)$$

where Y is the dependent variable, x_1 and x_2 are independent variables, β_1 and β_2 are unknown parameters, e is the error.

A series of $n > 2$ measurements are made on Y corresponding to preselected values of x_1 and x_2. Then using the principle of least squares estimates of the unknowns β_1 and β_2 can be found. There are several ways of obtaining the so-called normal equations associated with equation (1). However, the following rules give a simple method without resorting to calculus techniques.

Rule 1
Write down the model without the error term and replace the parameters with their estimates.

Equation (1) becomes

$$Y = b_1 x_1 + b_2 x_2 \qquad (2)$$

b_1 is the estimate of β_1, b_2 is the estimate of β_2.

Rule 2
For each estimate in turn multiply both sides of the equation by its coefficient to form a series of equations.

Equation (2) thus generates

coefficient of b_1 is x_1 giving $x_1 Y = b_1 x_1^2 + b_2 x_1 x_2$

coefficient of b_2 is x_2 giving $x_2 Y = b_1 x_1 x_2 + b_2 x_2^2$

Rule 3
Sum all equations over the n measured observations. This gives the normal equations

$$\Sigma x_1 Y = b_1 \Sigma x_1^2 + b_2 \Sigma x_1 x_2$$

$$\Sigma x_2 Y = b_1 \Sigma x_1 x_2 + b_2 \Sigma x_2^2$$

In these equations the unknowns are b_1 and b_2. Given a set of data the remaining quantities are all known. Thus any of the standard techniques used for solving a set of simultaneous equations such as

$$7 = 2x + 3y$$

$$3 = x + y \qquad (3)$$

can now be used.

At this stage the problem of statistical estimation of the model is given in equation (1) and the numerical solution of simultaneous equations are identical.

However, by suitable modification of the model in equation (1) we can also solve the simultaneous equation problem given by (3) and thus use proprietary software for solving such systems.

Using MODISTAT

Firstly in mathematics the number of equations is generally equal to the number of unknowns and since the LHS will equal the RHS there is no need for the statistical error term. Thus the model becomes

$$Y = \beta_1 x_1 + \beta_2 x_2 \qquad (4)$$

Hence if we now let the dependent variable Y represent the LHS values of equation (3), independent variable x_1 represent the coefficients of x, independent variable x_2 represent the coefficients of y, β_1 represent the unknown x and β_2 represent the unknown y we have a one to one correspondence and a statistics package can now solve the simultaneous equation problem.

Example 1 To solve

$7 = 2x + 3y$

$3 = x + y$

using MODISTAT.

The MODISTAT data file shown below is set up

```
                           MODISTAT V1.07
              LHS    X           Y
      ROW 1    7     2           3
      ROW 2    3     1           1
To display the OPTION MENU please press Return.?
```

Save the data on disc and return to the procedure menu. Using option 14 we obtain the following menu

MODISTAT V1.07

; CODE	:	OPTION FOR	:
: ROW COL	:	MULTIPLE REGRESSION	:
: − 1	−	STANDARD MODEL WITH INTERCEPT	:
: − 2	−	THROUGH ORIGIN	:
: − 3	−	WEIGHTED ANALYSIS WITH INTERCEPT	:
: − 4	−	WEIGHTED ANALYSIS THROUGH ORIGIN	:
:	−	ADD 4 TO THE ABOVE CODES TO PRODUCE	:
:		THE ANALYSIS ON ANOTHER DATA SET	:
: − 9	−	EXIT TO ANOTHER PROCEDURE	:

Type the OPTION CODE then press Return.?

Select option 2 since our 'model' has an intercept of zero.

MODISTAT V1.07

INDEPENDENT VARIABLES

Type the NUMBER OF INDEPENDENT VARIABLES then press Return.?

In this example there are two 'independent' variables, the coefficients of x and the coefficients of y. These are also entered.

MODISTAT V1.07

DEPENDENT VARIABLE

Type the COLUMN NAME then press Return.?

This is the LHS of our equations stored in the column called LHS.

MODISTAT V1.07

CODE	OUTPUT OPTION
1	– FITTED MODEL
2	– ANALYSIS OF VARIANCE TABLE
3	– TABLE OF OBSERVED, PREDICTED AND RESIDUALS
4	– SAVE RESIDUALS ON A DATA FILE
5	– DELETE AN INDEPENDENT VARIABLE
6	– RE-ENTER AN INDEPENDENT VARIABLE
7	– MULTIPLE CORRELATION COEFFICIENT
8	– ANOTHER REGRESSION USING THIS DATA
9	– REGRESSION USING A NEW DATA SET
10	– EXIT TO ANOTHER PROCEDURE

Type the OPTION CODE then press Return.?

To obtain the solution to our equation select option 1.

MODISTAT V1.07

FITTED MODEL		
VARIABLE NAME	ESTIMATE	STANDARD ERROR
X	2	
Y	1	

To display the OPTION MENU please press Return.?

To show that the solution is correct select option 3.

MODISTAT V1.07

NAME	OBSERVED	PREDICTED	RESIDUAL
ROW 1	7	7	0
ROW 2	3	3	0

To display the OPTION MENU please press Return.?

Example 2   Consistent equations

To solve $7 = 2x + 3y$

$3 = x + y$

$5 = x + 3y$

The MODISTAT data files are set up as before; however, the output now produced is shown below:

MODISTAT V1.07

FITTED MODEL		
VARIABLE NAME	ESTIMATE	STANDARD ERROR
X	2	
Y	1	

To display the OPTION MENU please press Return.?

MODISTAT V1.07

NAME	OBSERVED	PREDICTED	RESIDUAL
EQ 1	7	7	0
EQ 2	3	3	0
EQ 3	5	5	0

To display the OPTION MENU please press Return.?

Notice that the residuals are zero showing that the equations are consistent.

Example 3   Inconsistency

To solve $7 = 2x + 3y$

$3 = x + y$

$5 = 2x + 3y$

The MODISTAT data files are set up as before; however, the output now produced is shown below

MODISTAT V1.07

FITTED MODEL

VARIABLE NAME	ESTIMATE	STANDARD ERROR
X	2.999999	4.358901
Y	7.748604E-07	3.000001

To display the OPTION MENU please press Return.?

MODISTAT V1.07

NAME	OBSERVED	PREDICTED	RESIDUAL
EQ 1	7	6.000001	0.9999995
EQ 2	3	3	0
EQ 3	3	6.000001	−1.000001

To display the OPTION MENU please press Return.?

Notice that the residuals are now non-zero showing that the equations are inconsistent.

2. THE USE OF GRAPHICS TO ILLUSTRATE MATHEMATICAL CONCEPTS

The eye is our most important communication channel able to assimilate vast amounts of information regarding shape and form very rapidly. Since most modern microcomputers support good quality colour graphics this facility should enable us to exploit the eye to give students a better understanding of mathematical concepts. All the graphs associated with the following illustrations were obtained using MODISTAT.

2.1 Envelopes

For example the function $y = x \sin x$ has as its envelope $y = x$ and $y = -x$.

This is illustrated in Fig. 1 together with the fact that the local maximum of $x \sin x$ do not occur at ...

$$\frac{-3\pi}{2}, \frac{-\pi}{2}, \frac{\pi}{2}, \frac{3\pi}{2} \ldots$$

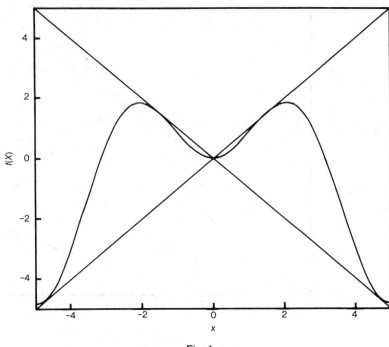

Fig. 1

2.2 A second example

Another interesting family of curves is illustrated by the following example

$y = 1 - \exp(-x)$

$y = 1 - \exp(-x^{1.5})$

$y = 1 - \exp(-x^2)$

These curves have several interesting properties in the range $x > 0$

— they are bounded between $y = 0$ and $y = 1$
— the curves are monotonic increasing
— at $x = 0$ $y = 0$
— the curves are continuously differentiable.

However, they have another interesting feature as is shown in Fig. 2.

The general family of these curves

$$y = 1 - \exp\left(-\frac{x}{\eta}\right)^{\beta}$$

are used extensively in the study of the reliability of systems. Providing η remains fixed (1 in our example) the family of curves always intersect at $x = \eta$ and is referred to as the characteristic life. The value of y corresoponding to $x = \eta$ is always $0.632 = 1 - e^{-1}$.

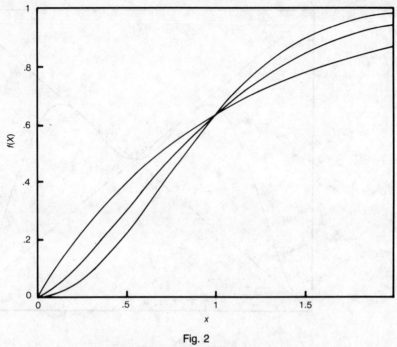

Fig. 2

2.3 Visual solution of equations and the remainder theorem

Consider the problem of finding the non-zero root of

$$e^x - 2x - 1 = 0$$

If we divide this equation into

$$y_1 = e^x \quad \text{and} \quad y_2 = 2x + 1$$

then the solution of the original equation will occur when $y_1 = y_2$.

The graphs are shown in Fig. 3. The equation has two roots at $x = 0$ and $x = 1.3$ (approximately).

It is of interest to note that $y_1 > y_2$ for $x < 0$; $y_1 < y_2$ for $0 < x < 1.3$ and $y_1 > y_2$ for $x > 1.3$.

This is of course the principle of the remainder therorem.

2.4 Fourier series

Fourier series are basically a method whereby certain functions can be represented as a series of sines and cosines.

We will be concerned with the convergence of the series rather than their derivation. In particular the Fourier expansion of x^2 is given by

$$x^2 = \frac{\pi^2}{3} + 4 \sum_{r=1}^{\infty} (-1)^r \frac{\cos rx}{r^2} \quad -\pi < x < \pi$$

The convergence of this series is shown in Fig. 4. The first graph shows plots of $y = x^2$, $y = \pi^2/3$ and the first harmonic $-4 \cos x$.

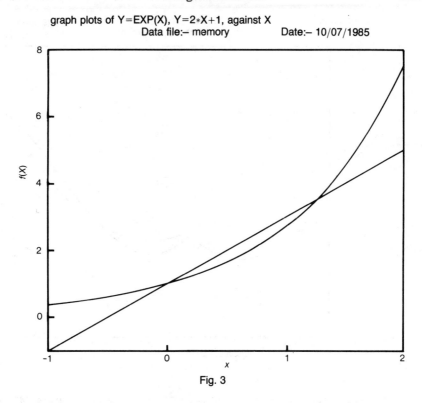

Fig. 3

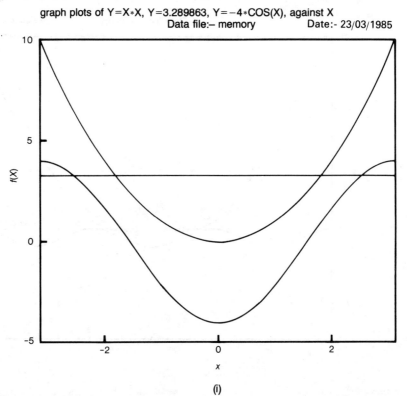

(i)

graph plots of Y=X∗X, Y=3.289863−4∗COS(X), Y=COS(2∗X), against X
Data file:− memory   Date:− 23/03/1985

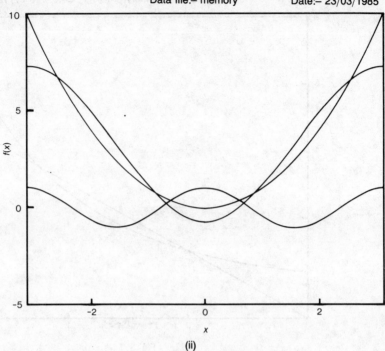

(ii)

graph plots of Y=X∗X, Y=3.289863−4∗COS(X)+COS(2∗X), Y=−0.4444445∗COS(3∗X), against X
Data file:− memory   Date:− 23/03/1985

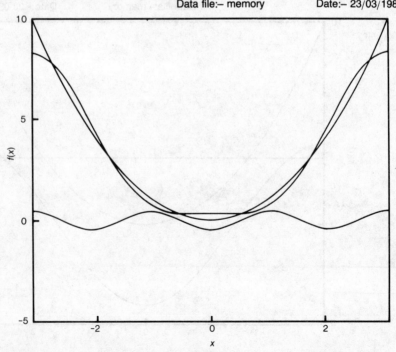

(iii)

graph plots of Y=X*X, Y=3.289863−4*COS(X)+COS(2*X)−0.4444445*COS(3*X).
against X
Data file:− memory   Date:− 23/03/1985

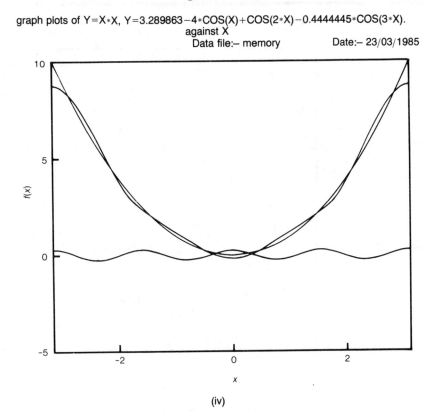

(iv)

Fig. 4

The second graph shows plots of $y = x^2$, $y = (\pi^2/3 - 4 \cos x$ and the second harmonic $y = \cos 2x$ etc.

As one can observe the convergence is very rapid apart from the extremes of the range. It is also interesting to note for successive harmonics how the amplitude decreases and the frequency increases.

Further Fig. 5 shows clearly that when $y = x^2$ is extended periodically it is continuous and therefore the series can be differentiated to produce the Fourier expansion of $2x$ or equivalently that of x. However, as the plot also shows the periodic extension of $y = x$ is not continuous and therefore the term by term differentiation of the associated Fourier series is not valid.

2.5 Discontinuities, limits and the mean value theorem

A lot of interest centres around the behaviour of functions at points of discontinuity. Figure 6 shows the plot of the function $(x^3 + 4x + 6)/(x^2 - 6x + 8)$ showing the discontinuities at $x = -4$ and $x = 2$.

Further, Fig. 7 shows the plot of the graphs

$$y = \sin x \quad y = \frac{1}{x} \quad \text{and} \quad y = \sin x / x$$

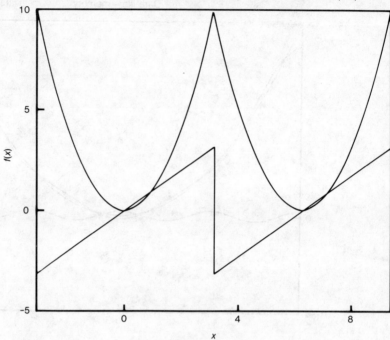

Fig. 5

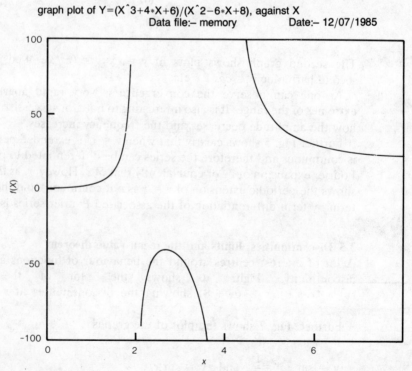

Fig. 6

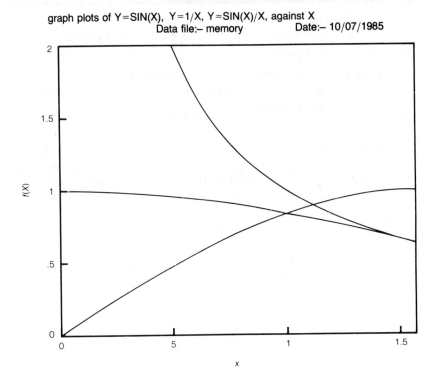

Fig. 7

It shows that

$$\lim_{x \to 0} \frac{\sin x}{x} = 1$$

and also illustrates the results of the mean value theorem

$$1 > \frac{\sin x}{x} > \frac{2}{\pi} \qquad 0 < x < \frac{\pi}{2}$$

3. CONCLUDING REMARKS

MODISTAT's function plotting routine allows a variety of functions to be investigated simply by typing in the function as a legal BASIC statement. Further, by using the IF statement step function and periodic extensions to functions can be plotted.

e.g. Y = X*X : IF X > 3.14159 THEN Y = (X − 6.28318) ↑ 2.

gives the periodic extension of $y = x^2$ used in Fourier example.

Due to MODISTAT's ability to cope with values outside the numeric range of the computer, functions with discontinuities can be easily investigated. For teaching purposes hard copies can be obtained in a variety of ways:

- Photographed using a single lens reflex camera to produce 35 mm transparencies.

- Output to large range of dot matrix printers.
- Output to flat bed plotters to produce quality graphics or overhead transparencies.

I hope that this chapter has produced some interesting ideas on the way in which computer graphics and proprietary software can be used to illustrate mathematical concepts.

REFERENCES

Draper & Smith. (1981). *Applied Regression Analysis*, Wiley.
Mangles, T. H. (1984). *MODISTAT Users Manual*, MODICOM.
Stephenson, G. (1973). *Mathematical Methods for Science Students*, Longmans.

23

'Soft' Course Simulation

D. E. Prior
Sunderland Polytechnic

and

R. S. Saunders
Manchester Polytechnic

---

1. INTRODUCTION

The past few years has seen a tremendous upsurge in the use of computers in all sections of society. The development of cheap but powerful computing facilities, especially the microcomputer, has brought about a demand for computing knowledge in areas far removed from those traditionally associated with calculating machinery. Now, not only science and engineering make their needs known in the demand for people skilled in using computers, the arts also seek experienced users of computers to assist in their studies of an ever more complex world. Such needs are reflected in current courses covering many disciplines at both undergraduate and postgraduate level by the inclusion of elements of computing and computer studies. Traditionally such course developments have tended to assume that the students on the receiving end of a computing input to their chosen course will in all cases benefit from a mathematical treatment of the subject and that to this end the students are competent mathematically. This assumed necessary predeliction for things mathematical is an unfair and quite untenable premise to adopt as a basis for the inclusion of computer studies and computing in courses which may take as their input a mixture from both the traditional streams of A level studies.

First, it may be a matter of student choice, for it is a curious reflection in the waters of time that mathematics, the most 'artistic' of academic pursuits, is viewed quite definitely as companion subject only to the 'hard' sciences. Ask students of the arts why they have chosen to pursue their particular line of study and in many cases the response will be 'well, I never was much good at maths. . . .' or 'I don't have a mathematical mind . . .'.

Secondly, and this is a reflection upon the training of teachers in the use of the computer in the classroom, many children are still encouraged to treat the computer as an extension of the mathematics equipment, though the mass media effectively often demonstrate that this is not the case.

Taking the above points together, it is no surprise therefore when, in the throes of an arts based course, students, upon finding themselves face to face with lecture material from the departments of mathematics, computer studies or computing, are heard to question 'why?' and query its relevance to their chosen fields of study.

In the softer disciplines of, for example, business, politics, sociology, commerce and environmental studies, it is not difficult to establish the relevance of a knowledge of computers and computing to each of these. It seems entirely natural that there should be a growing demand for computer involvement in soft courses as course leaders realise the impact of the computer revolution on their specialisms and seek to strike a balance between the traditional needs of the discipline and the demand to keep pace with developments in technique and methodology. The point is that though staff are sure of it students often fail to see the relevance of the computing and any mathematics that they do to the needs of their subject areas.

The last phrase above '. . . that they do . . .' is a key one for haste is made to add that the fault for the currently observed acceptance levels of mathematics and computing in many courses by no means wholly lies in the preparatory stages of students for further and higher education. New Polytechnic and University courses are still mounted (or old ones revised) in which computing, following the suggestions of staff members feeling 'something in the wind', is introduced to bring the courses up to date. Until very recently little constructive research was carried out by either serviced or servicing departments to establish how, in what way and at what price computers were affecting the serviced discipline. Syllabuses were lifted, with only minor, if any, modification from standard introductory material or self-contained units of existing computing courses. The computing department may have been a subsidiary of the 'Queen and Servant' department herself, yet little cross-referencing to the mathematics service syllabus supplied for that very same course occurred. In many cases the service mathematics syllabus could be found to be that drawn from standard introductory material to courses based solely in the mathematics discipline.

All of the above can add up to students taking a degree in a subject where though, to them, computers and computing clearly are playing an ever increasingly important part in the subject at large their feelings are quite strong that the computing and mathematics elements of the course are irrelevant, pointless exercises in equation and number manipulation.

To help combat this attitude requires a two-pronged approach. On the one hand, today's students must reawaken to the old-fashioned precept that the courses they choose are designed to be academically stimulating and demanding, as well as vocationally oriented. Many an erstwhile radical bent upon applying the pruning knife of relevance to some tree of knowledge in the forests of academe would do well to consider the CNAA documents concerning achievement of academic aims and objectives. The work students do as undergraduates should fire curiosity, fuel interest and feed the mind, not just sow the seeds for a harvest of pay-packets. This important aspect of education seems forgotten on the treadmill of qualification gathering in order to reduce the likelihood of unemployment.

On the other hand, the academic preparing for service courses should make the effort to determine to what extent, and in what ways, his subject is involved in the works which draw their recruits from the ranks of the serviced courses and devise his material accordingly. Nowhere is this more true than in mathematics, computing and computer studies where the extra-departmental service load must surely stand on a par with that for the department's own courses. And in these days in which many departments cast their eyes over work normally farmed out to the experts, a servicing department may run the risk of not being invited back because the serviced course leader feels that a better and more satisfying (to the student) job can be done from the inside where the relevance aspects are more clearly seen, understood and emphasised.

In recent years the involvement of the staff of the Mathematics and Computer Studies Department of Sunderland Polytechnic in discussions centred upon the revision of existing and the provisioning of new syllabuses for its courses both home based and serviced has caused much soul searching.

The place and role of the mathematics, computing and computer studies taught has been questioned and examined in the light of the aspirations of the courses concerned as reflected in their vocational and academic aims and objectives.

It must be said straight away that this chapter does not set out to consider points arising from the question of mathematics or computing and computer studies in the broader context, only those aspects which impinge upon modelling and simulation.

So far as modelling and computer-based simulation is concerned it was very soon realised that no course serviced did not boast some literature covering those aspects of the subject wherein the computer was being used as a vehicle to carry both predictive and hypothesis testing models. The main trouble was, of course, that links from these new sources to the undergraduate were not easily forgeable because of the need to develop considerable analytic machinery first. For example the modelling of growth and development for biologists requires considerable knowledge of set theory, computability and the Chomsky languages, and the environmental systems studies of Bennet and Chorley calls for a sound understanding of control theory. Usually, in order to be able to study a model which is not trivial the student is required to master some complex analytic technique which, of course, immediately opens the door to the problem of non-relevance and boredom sets in before enough can be mastered to build links to the subject area. In order to circumvent this problem past efforts have been to redraft the qualitative models encountered in terms of tally systems and elementary statistical analyses of distributions arising from tallying. But such models can lack interest and realism to some students. Inside 'soft' courses experiences are judged by driving word models of the experience through the processes of validation, experimentation and comparison with their external real world counterparts, and though the outcome of such processes must be relative to the value judgements of the participants, the pleasure of involvement for the student is great and under good tutoring, critical faculties are developed, insight occurs and learning takes place. Any quantification aspects often reside in the aforementioned statistical domain and act only as vindication of the acceptance or rejection of a hypothesis put forward in debate.

In the light of all the above, it seemed that some form of modelling and computer-based simulation was a possible way to answer the need to foster a quantitative approach in the 'softer' disciplines and both prepare the way for new technologies and encourage courses to face new, developing techniques.

The problems to be solved were 'what sort of modelling and what sort of simulation?'

The philosophy of modelling and the art and technique of model construction is, at this very moment, undergoing considerable re-examination. It is becoming realised that the origins of a model depend more upon the solution techniques available than was at first believed. A great deal of lip serivice is paid to the art of model building—to the need to establish clear goals and points of perception both with regard to the model itself and to the requirements of the modeller. But when all this has been voiced it is soon forgotten in the haste to strike out, through carefully reasoned argument, of course, this or that embarrassing term and quickly get to that differential equation. From then on 'modelling' is reduced to a process of obtaining the necessary tools, techniques and facilities to solve the equation. Clearly this approach is not desirable bearing in mind the purposes for which the topic is introduced into the scheme in the first place. Often the beginning modeller can be put off by too much techniques learning to start with. Indeed, has it not been discussed at length that this very style of approach probably caused much of the trouble in the first place.

To seek to establish any sort of mathematical relation between arbitrarily chosen variables very early on in the modelling scheme is not advisable on a serviced course. Most models in the 'softer' disciplines constructed along these lines for undergraduate assimilation tend to be so trivial that they offer nothing alongside the complex issues which can arise and be tackled by setting up verbal models and thrashing them out in a debate. As stated before any more worthwhile model demands knowledge of some part of the mathematical modeller's analytical armoury.

In addition, in order to satisfy the desire to introduce the student to computers and computing in their subject, the model ideally should lend itself for computer solution without making it appear that the computer is being used like a sledge-hammer to crack a nut. Most mathematical models do not need a computer to solve them, indeed, a part of the art of mathematical modelling is the study of expressions to prise the secrets of their behaviour from the form of the expression alone.

Whatever was decided upon the following points had to be met namely, that the approach

- could set up sufficiently complex models to ensure that the contrast with the traditionally qualitative approach, and all that that entailed, of a 'soft' subject was not naked and trivial;
- was clearly relevant to and part answered the developing needs of the subject area as expressed through the involvement of computers and computing in the field of work which looked to that subject for its fresh or graduate employees;
- was capable of supporting fairly complex models whose analytic tractability did not depend upon having to exhaust enthusiasm by learning new techniques applicable solely to solving the model;
- provided ample computer awareness and opportunity for beneficial computer–student interaction;
- was academically stimulating;
- supported other topic areas of the course.

Out of a number of alternative computer-based modelling and simulation methodologies an approach was found which answered most of the above points. This approach is founded on modelling using causal loop diagramming and then simulation through difference equation construction in a beginner's programming language. The option is also present for the student to translate the model into a form acceptable to the DYNAMO compiler if this option should be of benefit to the course.

As is demonstrated in the following example casual loop diagramming is a simple, but powerful, tool applicable to all disciplines and which immediately gets the students modelling in an exploring and critical manner. The causal loop diagram lends itself naturally to the establishing of difference equations, and it is soon realised by the quicker student that the use of the table functions 'softens' the models sufficiently to make them a viable adjunct and accessory to the more traditional forms of analysis in their subject. The use of the computer encompasses having to get a difference equation model up and running and where required, extends to using DYNAMO, a piece of commercially supplied simulation software.

The MSc. Mathematical Modelling and Computer-Based Simulation, MSc. Ecology, BSc. Environmental Studies, BSc. Joint Scheme of Sciences Combined Studies in Science and BA Data Processing degrees at Sunderland Polytechnic all include, along with mathematical modelling, modelling and simulation based upon causal loop analysis.

In order to demonstrate the method a class exercise for use in environmental studies is described below.

2. GROSS WORLD MODEL

2.1 Discussion

This globe of ours is of finite size. The amount of fertile land on it is not infinite but has measurable extents. It is not unreasonable therefore to assume that there is a ceiling to the maximum amount of food which the fertile land of Earth can support by natural growth and replenishment at any one time.

The peoples of the Earth have ever open mouths. As a result of the constant consumption of food the growth ceiling is never reached. At any point in time the total amount of food being replenished by the fertile areas of the globe is always less than the growth ceiling.

Environmentalists, contemplating the ultimate fate of a world of finite capacities being dwelt upon by an ever increasing number of inhabitants, have sought ways of assessing the consequences. Qualitative discussion grants that such growth cannot proceed indefinitely but only hints at the probable end and prevaricates about the route taken there.

As a first attempt towards some crude quantification of the problem a group of environmentalists have considered the above and put forward three indicators of the state of the global food condition. The first indicator is the gap between the food ceiling and the actual amount of food growing at any point in time. The second indicator is the ratio of the actual food growing to the food ceiling. The third indicator, called the gap regeneration time, is the amount of time, starting from a given point in time, needed by the Earth to close completely the food gap existing at that given point in time *provided nothing on Earth ate another morsel of food*.

The environmentalists further declared that two irrefutable statements could be made concerning the state of the global food supply in terms of the three factors described above.

(i) At any point in time the amount of food being regenerated through natural growth is directly dependent upon the food gap.
(ii) The gap regeneration time increases more and more quickly as the food ration decreases.

Discussions with biologists and demographers gave the environmentalists reasons to propose that a further important factor to consider was that of crowding—the extent to which each member of the human race encroached upon the living space of his neighbours. It was decided that the effects of crowding generally would be to lower fertility and shorten the average life-span of humans. In short, as the Earth becomes more and more crowded the incidence of live births per thousand of the population per year would become increasingly depressed and the average life-span of the total population would gradually fall.

On the other hand though, as biologists were quick to point out, there is nothing like a full stomach for peace of mind and encouraging the satisfaction of other appetites and so when the food consumed per capita is high it may be expected that the live births per thousand per year would be encouraged.

Discuss the above statement and draft a causal loop model which may be used to support ideas put forward concerning the future of man on a finite globe with finite food resources.

As each element is committed to the model briefly consider, where possible, the sources of data and information about it confirming that some reasonable estimate of its type and magnitude is obtainable from the sources at your disposal without too much difficulty.

2.2 Model construction: causal loop diagramming

One of the reasons why causal loop diagramming is proving popular in the 'softer' disciplines is because the student does not lose sight of his subject for a moment while constructing the model. At each step in establishing the model the problems to be solved are those which are common to the subject area and not introduced as a result of the techniques ultimately needed to 'solve' the model. The next step, which follows the instructions of the discussion document, illustrates this point.

Element commissioning is done through group discussion and considers, *in the light of the problem description only*, what things to include in the model, in what units they are to be measured and how information about them is to be obtained. The italics above emphasise the major importance of this statement. Construction of a model is *never* attempted without first having produced a statement embodying either the goal, aim, designed function or statement of action of the model. No matter how long, or troublesome, may the effort to produce such a statement be, it *is* produced and all else devolves from it, or derivations of it arising from amendments to it carried out as feedback from various ensuing stages occurs.

The following list of elements is typical of that eventually obtained from second year student discussions of the Gross World model.

Appendix 1 illustrates a possible causal loop diagram based upon the above elements.

Briefly, construction of the model is as follows. (For a fuller exposition of the technique of causal loop model building attention is drawn to the references at the end of this chapter.)

Each pair of elements of the model is examined in turn to see if either one of the pair affects the other in some way. The question of 'affects' is settled by asking 'if the chosen measure of this element changes (increases or decreases) then does the chosen measure of that element change (increase or decrease) due to the influence of this element over that one?' If the causal link is believed to exist then this link is established in the model by drawing an arrowed line from the influencing to the influenced element and writing a positive or negative sign close to the head of the arrow. The type of sign signifies the direction of the influence at work: *positive* if the change in measure of the influenced element is *in the 'same' direction as the change in that of the influencing element* and *negative*

308　Mathematical Modelling—Methodology, Models and Micros

Gross World Model—Elements Lists

Elements and units	Information source
• Maximum population of the Earth People	Research, discussion, guesswork
• Population of the Earth People	Departmental data
• Population ratio Dimensionless	Population/maximum population
• Global birth-rate Fraction of population/year	Departmental data and research to determine trend with crowding
• Global death-rate Fraction of population/year	Departmental data and research to determine trend of average life-span with crowding
• Global habitable space Square units	Research, discussion, guesswork
• Dynamic food ceiling of the Earth Food units	Departmental discussion, biologists, ecologists
• Current dynamic food level of the Earth Food units	Departmental discussion, biologists, ecologists
• Average normal global food needs Food units/person/year	Biologists
• Food gap Food units	Food ceiling-Dynamic food level
• Food ratio Dimensionless	Dynamic food level/food ceiling
• Food regeneration rate Fraction of food gap/year Food units/year	Food gap/gap regeneration time

if the change in measure of the influenced element is *in the 'opposite' direction to the change in that of the influencing element*.

The model is completed by tracing out all closed loops made up of successions of influence links and entering on the diagram the nature of the feedback at work in the model arising from that particular loop, positive for a growth-promoting loop and negative for an equilibrium-seeking loop.

This last activity, called 'model loop analysis', is of the utmost importance and some time is spent in ensuring that the principles of feedback in human involved, and human designed, systems is appreciated. Most students are aware of feedback and its effects and in the case of Environmental Studies students the principles and effects of it are discussed qualitatively and quantitatively in course topics from pollution to demography.

All of the above processes is surely modelling in the true sense of the term. A verbal model of an experience has been explored in debate and from it and the discussion certain elements, considered important in the light of the aims of the modeller and of the model itself, abstracted. By the application of a simple technique coupled with the exploration of course connected topics further elaboration and study of the experience is obtained to yield further information about, and an enlarged understanding of, the model. No clouding of issues has arisen through the need to force upon an unwilling audience the removal of elements once believed pertinent to the problem because by not so doing the model would eventually prove intractable. No clouding of issues has arisen through the need to trivialise an interesting model in order to shoe-horn it into a pair of differential equations. And no 'fair model' is lost through faint hearts pursuant to attempts to grasp the significance of answers found through the blind application, under intensive tutoring, of standard solution techniques to linear and nonlinear equations and differential functions.

The causal loop model of Appendix 1 is complex. Any doubts as to the truth of the assertion of complexity should be removed by attempting to model the Gross World, with all five nonlinear dependencies, using traditional mathematical modelling techniques.

The next stage is the investigation of the structure of the model sector by sector. A model sector is that part of the model containing the complete flow of a single element which is subject to the law of the conservation of matter (conserved flow) together with all the information flows (non-conserved flow) which control the flow of the conserved element across the model boundaries. The Gross World model has two sectors corresponding to the two conserved flows of population and food. A sector's 'structure' is the totality of feedback loops which act together to give rise to the observed semi-permanent behaviour of the sector as evidenced by the change over time of some selected element in the sector. The structure of the model itself arises from the feedback loops that collectively give rise to its semi-permanent behaviour observed in the changes over time in that element which, in the opinion of the modellers as expressed in the problem statement, represents the output from the model.

This sector by sector investigation of the model by the student group, though, at this stage of the modelling exercise, still essentially qualitative in treatment, often results in a much deeper and broader understanding of the topics surrounding, and upon which is based, the model. This seems to stem from a feeling of confidence in the model gained from its having been constructed, with justification, from the subject area itself and not thrust upon the students from another subject area, carrying with it its rules, regulations and seemingly arbitrary restrictions.

2.3 The computer model

Appendix 2 displays output from a microcomputer program designed to act as an environment in which models similar to the Gross World model, having been appropriately programmed from their causal loop counterparts may be driven through simulated time and their behaviour displayed as a series of trajectories of selected elements in the model. To reach this stage of the analysis the course may take one of two routes depending upon the objectives of the course for which the modelling and simulation exercise is being conducted. One route stems from a declared objective of the course being that the student should be capable of writing a computer program in a high level language and executing it. The other route stems from a course objective which declares that the student should be familiar with, and be competent in the use of, a commercially produced piece of computer software and have had exposure to computers through use of their operating systems to obtain answers to problems. The first route invariably includes the second, but not always does following the second route cover the first.

Most of the courses at Sunderland which contain modelling and simulation, either of the style of modelling described here ('systems' modelling) or of the traditional school of mathematical modelling, carry also a course in a high level programming language and so the Sunderland Environmental Studies students draft models in both a current high level general purpose programming language and in a form suitable for input to DYNAMO—a MIT produced compiler usd to execute, and poduce output from, models derived from causal loop diagrams along the lines described above. The DYNAMO compiler is extensively covered by texts mentioned in the references.

Appendix 2 is not obtained from the DYNAMO package, but from output produced by a Sunderland microcomputer program.

To illustrate briefly how Appendix 2 is obtained it is supposed, then, that the next step follows the Sunderland courses and is the conversion of the causal loop model into a set of general purpose programming language statements. Before the statements are produced by the students it is usually found that the use of a symbolic modelling stage interposed between the writing of the model program statements and the causal loop analysis is of great benefit, particularly if the model is very complex. This intermediate model, referred to as the 'system dynamics' model, is not entirely necessary, but can act as a check against inconsistencies in the model linkages. A system dynamics model takes the essentially qualitative, high level causal loop model and shifts it down the conceptual scale such that the mechanical task of computer model statement generation is only a matter of following the system dynamics model through each sector. For readers familiar with the activities of commercial system analysts the production of a set of computer statements from a systems dynamics model might be paralleled with that of coding a commercial program from the detailed flow-chart of the problem. The causal loop model may be paralleled with the analyst's system specification.

System dynamics model building was developed by J. W. Forrester of MIT from the symbolic block diagramming conventions of the control engineers. Again, for further amplification of this and related topics the interested reader is encouraged to explore the texts in the accompanying references.

Generating the models in a high level language, once a good grasp of the language has been obtained, is simplicity itself for the models are based upon the difference equation. As an example, take the part population sector concerning population, fertility, average life-span and birth- and death-rates. The statements for this section of the sector are as follows

```
FOR I=1 TO SIMLENGTH
        POP=POP+DT*(BIRTHS-DEATHS)
        CROWDING=POP/HABSPACE
        POPRATIO=POP/POPMAX
        GOSUB fertility(CROWDING,FFACT)
        GOSUB lifespan(POPRATIO,LFACT)
            BIRTHS=FFACT*POP
            DEATHS=LFACT*POP
NEXT I
```

fertility: returns FFACT by interpolation in the table in Fig. 1 below.
life-span: returns LFACT by interpolation in table similar to that in Fig. 1.

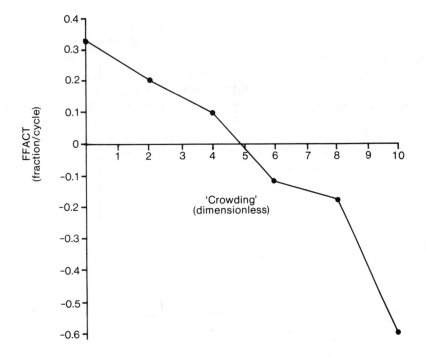

Fig. 1. Table relating effect of crowding on fertility factor.

Note DT is the small interval of time over which the rates of *births* and *deaths* per year are assumed constant. For the purposes of the Gross World model a value of 0.1 cycle is amply small enough.

Appendix 2 illustrates output from a series of simulation experiments carried out on the Gross World model. This form of modelling is useful for environmentalists because the graphical output both complements and reinforces the way in which other topics on the course both study and handle information. Take, for example, Fig. 1, a study of the behaviour of the small section of the model consisting of population subject to a fixed positive growth rate. First year students soon can produce this curve and as a result when the expotential growth curve is met either in demographic studies or in mathematical modelling understanding is enhanced through familiarity with the mechanisms and system structure which produces this effect. In addition, since each sector or part sector may be studied separately from the rest of the model, the underlying structure of a complex model may be mapped out gradually to explain which loops predominate and how the final structure of the total sytem may be related to the fights for loop dominance which take place in the dynamic arena of the living model as it is driven through simulated time.

Add to all this the fact that, first, as Appendix 2 most eloquently demonstrates, this form of modelling, as early as the second year, allows the non-scientific student the opportunity to compare and contrast a sympathetic form of scientifically oriented analytical methodology with those analytical tools of his chosen discipline: that secondly, the technique frees the student from the fetters of mathematical method, but may be used, if desired, as a confidence building preliminary to the discussion of mathematically based solution techniques or to complement results obtained through application of the traditional analyses of mathematical modelling: and that thirdly, since the technique upholds in its model building philosophy all that is desirable in the model origination stage then surely it must be agreed that one has a most useful, thought provoking, flexible and generally applicable, both in the sciences and in the arts, modelling and simulation methodology.

APPENDIX 1: GROSS WORLD MODEL CAUSAL LOOP DIAGRAM

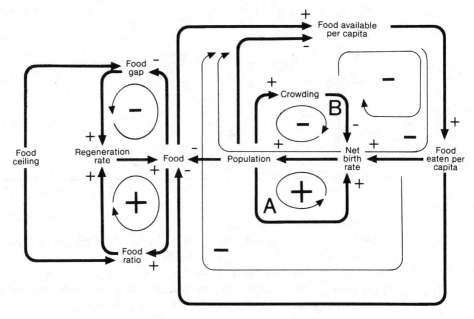

Appendix 1: Gross world model causal loop diagram

APPENDIX 2

Figure A.1 is a study of the effect upon the population of the single positive feedback loop labelled A in Appendix 1. The net birth rate initially is assumed to be positive. As expected the population trajectory is the familiar exponential growth curve.

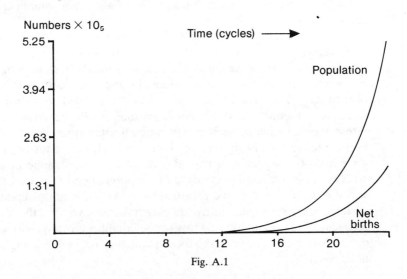

Fig. A.1

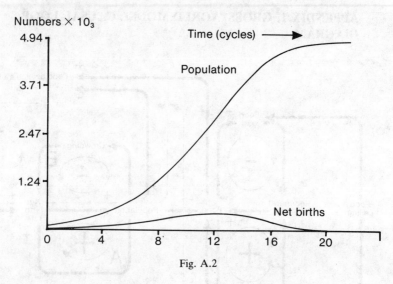

Fig. A.2

The results of expanding the study to include the effects of crowding are shown in Fig. A.2. The negative feedback loop labelled B in Appendix 1 is weak for low population values and loop A dominates giving exponential growth in the early stages. As crowding bites, exponential growth is moderated by the ever strengthening negative feedback giving near linear growth in the middle cycles. Finally, the presence of a finite amount of living space results in loop B dominance and the systems coasts asymptotically to its equilibrium value.

Again, the curve is a familiar one—the S-shaped, or sigmoidal growth curve.

In the above way the effects of any combination of feedback loops upon segments and sectors of the system can be studied. The ready graphical nature of the output assists in the demonstration and discussion of such important feedback system concepts as phase, gain, amplification and attenuation.

Once the whole model has been programmed the modeller is then encouraged to experiment. Before doing so, however, it is sometimes necessary to seek the steady state conditions for the system in order to be fairly sure that any transient response observed pursuant to an impressed disturbance has arisen from the disturbance and not from system settlement. Figures 3 and 4 below, for example, illustrate two runs of the Gross World model in each of which equilibrium running for the first six cycles prefaces a disturbance of some kind. The trajectories of Fig. A.3 arise from the supposition that as a result of a catastrophe in about the sixth cycle a substantial percentage of the population of the world has been wiped out. The scale of the catastrophe, however, is not so great that the system cannot recover fairly quickly as is shown by the asymptotic recovery of the population figures to the equilibrium value within six time cycles. The total change in the food quantity is negligible being about 0.1 units overall.

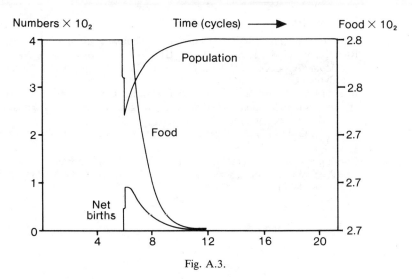

Fig. A.3.

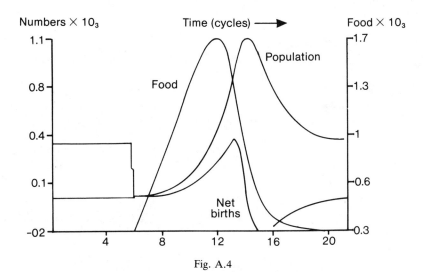

Fig. A.4

Figure A.4 tells the history of the system after a catastrophe which removes over 90% of the population in a short space of time. The trajectories are quite different from those of Fig. A.3 but readers familiar with system response characteristics will recognise the behaviour of a complex system undergoing exponential growth, inertial overshoot and collapse.

Once 'soft course' students have grasped the significance of this form of graphical output in relation to the feedback studies of the causal loop model of the real world system and that they can play God in a microcosm then in many cases the seeds are sown for a ready willingness to concede that the arts and sciences can harmonise in undergraduate courses to the benefit of each.

REFERENCES

Roberts, N. *et al.* (1983) *Introduction to Computer Simulation*, Addison-Wesley.

Richardson, G. P. & Pugh, A. L., III (1981). *Introduction to System Dynamics Modelling with DYNAMO*. MIT Press/Wright-Allen series in system dynamics.

International Conference on System Dynamics, Geilo, Norway (1976). *Elements of the System Dynamics Method*, (ed.) Jorgen Randers, MIT Press/Wright-Allen series in system dynamics. 1980.

Coyle, R. G. (1977). *Management System Dynamics*, John Wiley. Reprinted May 1978.

Goodman, M. R. (1974). *Study Notes in System Dynamics*, MIT Press/Wright-Allen series in system dynamics. Second printing by the MIT Press, 1980.

Index

A
ACSL
  illustrations of, 227
  problems with, 236
assessment, 48, 231

B
Bajpai, A.C., 48, 61
BCSSP
  simulation system, 238, 246
  model for connecting rod problem, 244
Bode plot, 209
book of stamps, 12
Bruner's theory, 268
Burghes, D.N., 11

C
case studies
  cam design, 200
  crankshaft torque, 199
  discussion of, 179
  marking of, 174
  non linear damping, 199
  use of, 192
  vehicle seat, 167
causal loop
  analysis, 305
  diagramming, 307
  model, 87
Centaur system, 280
circuit diagram, 13
cognitive science, 263, 269
computer-aided draughting, 224
computer-aided learning, 191
computer packages
  advantages in teaching, 227
  applications in engineering, 224
concept matrix, 18

connecting rod acceleration problem, 241
control, 273
'control aids' package for the BBC micro, 211
control theory, 205
  multi-input multi-output, 206
  single-input single-output, 206
CSMP simulation package, 255
  use in diffusion process, 256

D
data scramblers, 160
discrete linear systems, 155
DSA, 102
DYNAMO, 305

E
education data, 27
expert systems, 274

F
Faustmann formula, 183
feature list, 81
feedback control startegy, 206, 208
'feeding the world' problem, 306
  computer model for, 309
formulation-solution, 61
friction, 120

G
GCSE assessment objective, 15
GPS program, 270
group projects, 90
Guthrie's contiguous conditioning theory, 264

Index

H
Hall's behaviouristic reinforcement theory, 265
Hanna, G., 27
heuristics, 75
HND Maths, Stats & Computing, 90
Hubble's law, 143

I
iconic simulation, 224
information systems, 149
interpretation model, 283

K
Kaiser-Messmer, G., 36

L
Laplace transform method, 207
least squares, 137
Lewin's topological theory, 265
linear system, 152
LISREL program, 27
logic systems, 281

M
marking scheme, 95
mechanics, 116
model formulation, 80
model loop analysis, 308
modelling
 formulation stage by A.I., 263
 levels of, 240
 population growth, 250
 relevance of cognitive science to, 270
modelling process, 62
models
 cosmological, 144
 discrete, 130
 forest management, 181
 population growth, 132
MODISTAT, 287
 use in demonstrating fourier series convergence, 294
 use in solving simultaneous equations, 287
 use to illustrate behaviour of functions, 297
 use to illustrate mathematical concepts, 292

N
Newtonian mechanics
 computer games in, 219
 simulation for, 218
 with a micro, 217
Nichol plot, 210
normal modes, 123

Nyquist plot, 209

O
Oke, K.E., 48, 61
orbital motion, 220

P
PAFEC
 illustrations of, 227
 problems with, 236
pattern recognition, 273
Poisson process, 183
population growth modelling, 250, 306
Prior, D.E., 80
psychology, 263

Q
QSD, 102

R
reading age, 17
relationship level graph, 50
rocket problem, 220
root locus, 210

S
search strategies, 282
signals, 151
simulation
 discrete event techniques, 249
 dynamic, 238
 iconic, 224
 in mechanics, 218, 219
 of continuous systems, 255
Slater, G.L., 90
software libraries, 193
 CALNAPS, 194
 NAG, 197
software packages, 119
stochastic difference equations, 187
system dynamics, 99
system methodology, 101
systems thinking, 98

T
Thorndyke's law of effect, 264
Tolman's expectancy theory, 266

V
validation, 116
VESPA, 85

W
Wertheimer's gestalt theory, 265
Wolstenholme, E.F., 98

Mathematics and its Applications
Series Editor: G. M. BELL, Professor of Mathematics, King's College (KQC), University of London

Artmann, B.	The Concept of Number*
Balcerzyk, S. & Joszefiak, T.	Commutative Rings*
Balcerzyk, S. & Joszefiak, T.	Noetherian and Krull Rings*
Baldock, G.R. & Bridgeman, T.	Mathematical Theory of Wave Motion
Ball, M.A.	Mathematics in the Social and Life Sciences: Theories, Models and Methods
de Barra, G.	Measure Theory and Integration
Bell, G.M. and Lavis, D.A.	Co-operative Phenomena in Lattice Models Vols. I & II*
Berkshire, F.H.	Mountain and Lee Waves
Berry, J.S., Burghes, D.N., Huntley, I.D., James, D.J.G. & Moscardini, A.O.	Teaching and Applying Mathematical Modelling
Burghes, D.N. & Borrie, M.	Modelling with Differential Equations
Burghes, D.N. & Downs, A.M.	Modern Introduction to Classical Mechanics and Control
Burghes, D.N. & Graham, A.	Introduction to Control Theory, including Optimal Control
Burghes, D.N., Huntley, I. & McDonald, J.	Applying Mathematics
Burghes, D.N. & Wood, A.D.	Mathematical Models in the Social, Management and Life Sciences
Butkovskiy, A.G.	Green's Functions and Transfer Functions Handbook
Butkovskiy, A.G.	Structural Theory of Distributed Systems
Cao, Z-Q., Kim, K.H. & Roush, F.W.	Incline Algebra and Applications
Chorlton, F.	Textbook of Dynamics, 2nd Edition
Chorlton, F.	Vector and Tensor Methods
Crapper, G.D.	Introduction to Water Waves
Cross, M. & Moscardini, A.O.	Learning the Art of Mathematical Modelling
Cullen, M.R.	Linear Models in Biology
Dunning-Davies, J.	Mathematical Methods for Mathematicians, Physical Scientists and Engineers
Eason, G., Coles, C.W. & Gettinby, G.	Mathematics and Statistics for the Bio-sciences
Exton, H.	Handbook of Hypergeometric Integrals
Exton, H.	Multiple Hypergeometric Functions and Applications
Exton, H.	q-Hypergeometric Functions and Applications
Faux, I.D. & Pratt, M.J.	Computational Geometry for Design and Manufacture
Firby, P.A. & Gardiner, C.F.	Surface Topology
Gardiner, C.F.	Modern Algebra
Gardiner, C.F.	Algebraic Structures: with Applications
Gasson, P.C.	Geometry of Spatial Forms
Goodbody, A.M.	Cartesian Tensors
Goult, R.J.	Applied Linear Algebra
Graham, A.	Kronecker Products and Matrix Calculus: with Applications
Graham, A.	Matrix Theory and Applications for Engineers and Mathematicians
Griffel, D.H.	Applied Functional Analysis
Griffel, D.H.	Linear Algebra*
Hanyga, A.	Mathematical Theory of Non-linear Elasticity
Harris, D.J.	Mathematics for Business, Management and Economics
Hoksins, R.F.	Generalised Functions
Hoskins, R.F.	Standard and Non-standard Analysis*
Hunter, S.C.	Mechanics of Continuous Media, 2nd (Revised) Edition
Huntley, I. & Johnson, R.M.	Linear and Nonlinear Differential Equations
Jaswon, M.A. & Rose, M.A.	Crystal Symmetry: The Theory of Colour Crystallography
Johnson, R.M.	Theory and Applications of Linear Differential and Difference Equations
Kim, K.H. & Roush, F.W.	Applied Abstract Algebra
Kosinski, W.	Field Singularities and Wave Analysis in Continuum Mechanics
Krishnamurthy, V.	Combinatorics: Theory and Applications
Lindfield, G. & Penny, J.E.T.	Microcomputers in Numerical Analysis
Lord, E.A. & Wilson, C.B.	The Mathematical Description of Shape and Form
Marichev, O.I.	Integral Transforms of Higher Transcendental Functions
Massey, B.S.	Measures in Science and Engineering
Meek, B.L. & Fairthorne, S.	Using Computers

Mikolas, M.	**Real Function and Orchogonal Series**
Moore, R.	**Computational Functional Analysis**
Müller-Pfeiffer, E.	**Spectral Theory of Ordinary Differential Operators**
Murphy, J.A. & McShane, B.	**Computation in Numerical Analysis***
Nonweiller, T.R.F.	**Computational Mathematics: An Introduction to Numerical Approximation**
Ogden, R.W.	**Non-linear Elastic Deformations**
Oldknow, A. & Smith, D.	**Learning Mathematics with Micros**
O'Neill, M.E. & Chorlton, F.	**Ideal and Incompressible Fluid Dynamics**
O'Neill, M.E. & Chorlton, F.	**Viscous and Compressible Fluid Dynamics***
Page, S. G.	**Mathematics: A Second Start**
Rankin, R.A.	**Modular Forms**
Ratschek, H. & Rokne, J.	**Computer Methods for the Range of Functions**
Scorer, R.S.	**Environmental Aerodynamics**
Smith, D.K.	**Network Optimisation Practice: A Computational Guide**
Srivastava, H.M. & Karlsson, P.W.	**Multiple Gaussian Hypergeometric Series**
Srivastava, H.M. & Manocha, H.L.	**A Treatise on Generating Functions**
Shivamoggi, B.K.	**Stability of Parallel Gas Flows***
Stirling, D.S.G.	**Mathematical Analysis***
Sweet, M.V.	**Algebra, Geometry and Trigonometry in Science, Engineering and Mathematics**
Temperley, H.N.V. & Trevena, D.H.	**Liquids and Their Properties**
Temperley, H.N.V.	**Graph Theory and Applications**
Thom, R.	**Mathematical Models of Morphogenesis**
Toth, G.	**Harmonic and Minimal Maps**
Townend, M. S.	**Mathematics in Sport**
Twizell, E.H.	**Computational Methods for Partial Differential Equations**
Wheeler, R.F.	**Rethinking Mathematical Concepts**
Willmore, T.J.	**Total Curvature in Riemannian Geometry**
Willmore, T.J. & Hitchin, N.	**Global Riemannian Geometry**
Wojtynski, W.	**Lie Groups and Lie Algebras***

Statistics and Operational Research
Editor: B. W. CONOLLY, Professor of Operational Research, Queen Mary College, University of London

Beaumont, G.P.	**Introductory Applied Probability**
Beaumont, G.P.	**Probability and Random Variables***
Conolly, B.W.	**Techniques in Operational Research: Vol. 1, Queueing Systems***
Conolly, B.W.	**Techniques in Operational Research: Vol. 2, Models, Search, Randomization**
Conolly, B.W.	**Lecture Notes in Queueing Systems**
French, S.	**Sequencing and Scheduling: Mathematics of the Job Shop**
French, S.	**Decision Theory: An Introduction to the Mathematics of Rationality**
Griffiths, P. & Hill, I.D.	**Applied Statistics Algorithms**
Hartley, R.	**Linear and Non-linear Programming**
Jolliffe, F.R.	**Survey Design and Analysis**
Jones, A.J.	**Game Theory**
Kemp, K.W.	**Dice, Data and Decisions: Introductory Statistics**
Oliveira-Pinto, F.	**Simulation Concepts in Mathematical Modelling***
Oliveira-Pinto, F. & Conolly, B.W.	**Applicable Mathematics of Non-physical Phenomena**
Schendel, U.	**Introduction to Numerical Methods for Parallel Computers**
Stoodley, K.D.C.	**Applied and Computational Statistics: A First Course**
Stoodley, K.D.C., Lewis, T. & Stainton, C.L.S.	**Applied Statistical Techniques**
Thomas, L.C.	**Games, Theory and Applications**
Whitehead, J.R.	**The Design and Analysis of Sequential Clinical Trials**

*In preparation

THE LIBRARY
ST. MARY'S COLLEGE OF MARYLAND
ST. MARY'S CITY, MARYLAND 20686